Lecture Notes in Mathematics

A collection of informal reports and seminars
Edited by A. Dold, Heidelberg and B. Eckmann, Zürich

Series: Mathematics Institute, University of Warwick
Adviser: D. B. A. Epstein

285

Pierre de la Harpe

University of Warwick, Coventry

Classical Banach-Lie Algebras and Banach-Lie Groups of Operators in Hilbert Space

Springer-Verlag
Berlin · Heidelberg · New York 1972

AMS Subject Classifications (1970): Primary: 17 B 65
Secondary: 22 E 65, 17 B 20, 17 B 45, 18 H 25, 22 E 60

ISBN 3-540-05984-9 Springer-Verlag Berlin · Heidelberg · New York
ISBN 0-387-05984-9 Springer-Verlag New York · Heidelberg · Berlin

Offsetdruck: Julius Beltz, Hemsbach/Bergstr.

<u>Contents</u>

INTRODUCTION

This work contains three parts of similar lengths.

Chapter I is devoted to the study of some infinite dimensional Lie algebras of linear operators.

Chapter II to that of Banach-Lie algebras and Banach-Lie groups related to them.

And Chapters III and IV to applications : infinite dimensional symmetric spaces, cohomology of the stable classical groups.

The main results are briefly described in sections 3 to 4 of this introduction. The first two sections of the introduction are respectively concerned with various examples of Banach-Lie groups, and with some indications about the general theory of Banach-Lie groups.

I have found it necessary and helpful to attempt and draw together the relevant literature. Some of the references in the bibliography do not appear in the text. Those of particular importance for our purpose are indicated by a * .

0.1.- Banach-Lie groups : examples

Banach-Lie algebras and Banach-Lie groups arise naturally in many different contexts.

The earliest work devoted to them seems to be one by Pérès (1919) a propos of integral equations (Delsarte : introduction to Chapter IV of [42]).

Groups of operators on Hilbert space

The first non trivial examples are provided by the general linear group $GL(\mathcal{H})$ of all invertible bounded operators on a Hilbert space $\mathcal{H}$ over $\mathbb{C}$ or $\mathbb{R}$ and by various of its subgroups : unitary, orthogonal, symplectic groups ... More generally, the groups of units and the group of unitary elements in any associative involutive Banach algebra with unit are Banach-Lie groups. An important example is given by the C*-algebra $L(\mathcal{H})/C(\mathcal{H})$, where $L(\mathcal{H})$ is the algebra of all bounded operators on $\mathcal{H}$ and where $C(\mathcal{H})$ is the ideal of compact operators; both the group of units and the group of unitary elements are then homotopically equivalent to the space of Fredholm operators on $\mathcal{H}$, hence are both classifying spaces for K-theory; for the relevance of these two groups in K-theory, see Atiyah-Singer [7] and Eells [55].

If $\mathcal{I}(\mathcal{H})$ is a Banach algebra and an ideal in the associative algebra $L(\mathcal{H})$, then the subgroup $GL(\mathcal{H};\mathcal{I})$ in $GL(\mathcal{H})$ consisting of those operators of the form $id_{\mathcal{H}} + X$ with $X \in \mathcal{I}(\mathcal{H})$ is a Banach-Lie group locally diffeomorphic to $\mathcal{I}(\mathcal{H})$. The standard examples are when $\mathcal{I}(\mathcal{H})$ is the ideal of compact operators $C(\mathcal{H})$ - in which case $GL(\mathcal{H};C)$ is the so-called Fredholm group of $\mathcal{H}$ considered by Delsarte [42] - when $\mathcal{I}(\mathcal{H})$ is the ideal of Hilbert-Schmidt operators $C_2(\mathcal{H})$ -

in which case the Lie algebra of $GL(\mathcal{H}; C_2)$ is one of the simple <u>L*-algebras</u> [153] - and when $\mathcal{I}(\mathcal{H})$ is the ideal of trace class operators $C_1(\mathcal{H})$ - in which case the relationship between cohomologies of the group and of its Lie algebra are particularly interesting (see our Chapter IV).

Here is a somewhat more sophisticated example : Let E be a separable real Banach space furnished with a Gaussian measure γ and let $\tilde{G}$ be the group of those invertible operators X on E such that i) γ and $X_*(\gamma)$ are in the same measure class; ii) the Radon-Nicodym derivative d_X of $X_*(\gamma)$ with respect to γ is continuous on E ; iii) $d_X{}^{-1}$ is continuous on E. Then an explicit formula is know for d_X when X is in the subgroup G of $\tilde{G}$ defined as follows : Let $E^* \to \mathcal{H} \to E$ be the abstract Wiener space defined by (E,γ) and let $\underline{g}$ be the Banach-Lie algebra of those operators Y on $\mathcal{H}$ which extend to an operator $Y' \in L(E,E^*)$ as indicated in the diagram

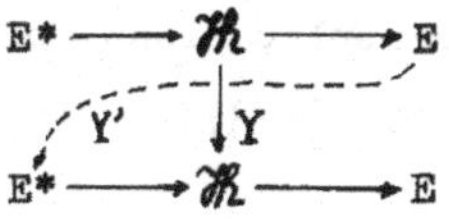

Then G is the sub Banach-Lie group of $GL(\mathcal{H}; C_1)$ with Lie algebra $\underline{g}$; and G is called the <u>Wiener group</u> of (E,γ). It does not seem to be known whether or not G is a proper subgroup of $\tilde{G}$. My references for this example are J. Radcliff and P. Stefan in [58].

Other examples of Banach-Lie groups acting linearly on vector spaces can be found in Kadison [93], [94], [95], Ouzilou [128], Rickart [138], [139], [140], Sunouchy [171].

Structural groups and manifolds of maps

As in finite dimensions, a large part of the current interest in infinite dimensional manifolds is devoted to extra structures on them, hence to the structural Banach-Lie groups of appropriate bundles. The orthogonal group of a real Hilbert space is the structural group of a Riemannian manifold (Eells [54] section 5). Problems partially motivated by theoretical mechanics have led A. Weinstein to a careful investigation of symplectic manifolds [182], [183]. Manifolds given together with a reduction of their structural group to the Fredholm group, called Fredholm manifolds, arise naturally in many concrete problems of global analysis : degree theory, elliptic problems; see Elworthy [61], Elworthy-Tromba [62], Eells [55], Eells-Elworthy [56]. The Wiener group is a basic ingredient for the theory of Wiener manifolds, recently developed by Eells-Elworthy [57], Eells [59].

Manifolds of maps themselves can be Banach-Lie groups. For example, let S be a compact manifold and let G be a finite dimensional Lie group. The space of those maps from S to G which belong to certain classes (continuous, or Sobolev if S and G have Riemannian structures) are naturally Banach-Lie groups under pointwise multiplication; they have good applications to the algebraic topology of homogeneous spaces of G (Eells [53]).

Automorphism groups of infinite dimensional geometric objects

The (historically) first example in this category is given by Wigner's theorem : Let $\mathcal{A}$ be the group of symmetries of a projective complex Hilbert space; more precisely, if $P(\mathcal{H})$ is the set of lines in the complex Hilbert space $\mathcal{H}$, the transition probability is defined by

$$\begin{cases} P(\mathcal{H}) \times P(\mathcal{H}) \longrightarrow \mathbb{R} \\ (\xi, \eta) \longmapsto \dfrac{|\langle X|Y\rangle|^2}{|X|^2|Y|^2} \end{cases}$$

where X [resp. Y] is any non-zero vector in the complex line ξ [resp. η]; a permutation of $P(\mathcal{H})$ belongs to $\mathcal{A}$ if and only if it preserves the transition probability. Let $\tilde{U}(\mathcal{H})$ be the Banach-Lie group of all unitary and antiunitary operators on $\mathcal{H}$; any element of $\tilde{U}(\mathcal{H})$ induces clearly a symmetry of $P(\mathcal{H})$ and there is a sequence.

$$\{e\} \longrightarrow U(1) \longrightarrow \tilde{U}(\mathcal{H}) \longrightarrow \mathcal{A} \longrightarrow \{e\}.$$

Wigner proved that this sequence is exact; hence $\mathcal{A}$ is a Banach-Lie group whose Lie algebra is the quotient of the Banach-Lie algebra $\underline{u}(\mathcal{H}) = \{X \in L(\mathcal{H}) \mid X^* = -X\}$ by its centre. A similar result holds for quaternionic spaces. Proofs and comments are given in detail by Bargman [18]. The short exact sequence written above is formally the same as that appearing in propositions I.10.A and II.3.A.

Let J be a complex Banach subspace of the complex C*-algebra $L(\mathcal{H})$ which contains X^* and X^2 whenever it contains the operator X, and let J(1) be the open unit ball in J. (In particular, if $J = L(\mathcal{H})$, J(1) is a generalization of the Siegel's generalized unit disc.) Then any holomorphic diffeomorphism of J(1) onto itself is the composition of a Möbius transformation and of a linear isometry; in particular this group of diffeomorphisms of J(1) is a Banach-Lie group. The proof of this statement uses a generalized Schwartz lemma and is due to L. Harris [73], [74].

Greenfield and Wallach [66], [67] have studied infinite dimensional analogues of the classical bounded domains of type I and have computed explicitely their groups of holomorphic diffeomorphisms. These groups are again Banach-Lie groups canonically given as subgroups of $GL(\mathcal{H})$, where $\mathcal{H}$ is a complex Hilbert space.

In a general situation, the following fact has been observed by Eells : Let M be an infinite dimensional Riemannian manifold which is connected and complete. Let G^+ be the connected component of its group of isometries. Then G^+ is a Banach-Lie group whose Lie algebra consists of the Killing vector fields of M. Indeed, an isometry of M is uniquely determined by its value and that of its first derivative at any one point of M; this finiteness condition implies the Banach character of the group G^+.

0.6

A result about Banach-Lie groups and finite dimensional manifolds

Almost all the previous examples are Banach-Lie groups which act on infinite dimensional spaces or manifolds. This happens to be a general fact, at least when some (rather weak) semi-simplicity hypothesis is assumed about the group. Indeed, the following result has been proved in [126] as a corollary of the theory of the primitive Lie algebras of E. Cartan.

Theorem 1. Let G be a connected Banach-Lie group and let M be a finite dimensional smooth manifold. Suppose there exists a smooth and effective action of G on M which is ample and primitive. Then G is finite dimensional.

Primitivity in theorem 1 means : If $\underline{g}$ is the Banach-Lie algebra of G, if G_o is the isotropy subgroup at some point of M and if $\underline{g_o}$ is its Lie algebra, then $\underline{g_o}$ is maximal among the proper closed subalgebras of $\underline{g}$. Any transitive action of a second countable Lie group is ample. (Details in [126].) A corollary of the proof is:

Theorem 2. Let G and M be as above and suppose that the Lie algebra of G contains no closed finite codimensional ideal. Then the only smooth action of G on M is the trivial action which associates to each element of G the identity transformation of M.

The groups introduced in Chapter II of the present work satisfy the conditions of theorem 2.

0.2.- Banach-Lie groups : general theory

There is a theory of local Banach-Lie groups (i.e. group-germs) and of Banach-Lie algebras, due to Birkhoff [20] and Dynkin [51], of which the results are essentially as in finite dimensions. The global theory, however, offers many facts without finite dimensional analogues. For instance, Banach-Lie algebras which are not enlargable, that is which are not Lie algebras of any Banach-Lie group, have been constructed by Van Est-Korthagen [177] and Douady-Lazard [49]. In contrast, there are good criteria, one of which is :

Theorem 3. Let $\underline{g}$ be a Banach-Lie algebra whose center is reduced to zero. Then there exists a Banach-Lie group of which $\underline{g}$ is the Lie algebra.

Theorem 3 will be good enough for our purpose; its proof can be found in Lazard [107], section 22. There are other criteria for enlargibility in [177] and Swierczkowski [173].

The best reference about the general theory of Banach-Lie groups is Lazard [107]; some of the technical definitions we need will be recalled in section II.1. The background about Banach manifolds can be found in Bourbaki [27], [28], Dieudonné [44], Lang [103], Lazard [106]. Other references about Banach-Lie groups include : Eells [54] section 3, Laugwitz [104], [105], Leslie [109], Maissen [116]; and about infinite dimensional manifolds : Kuiper [102], Moulis [194].

0.3.- Structure of the classical Banach-Lie algebras and groups of operators in Hilbert space

(Chapters I, II and IV.)

The primary purpose of this work is to study in detail certain classical Banach-Lie algebras and Banach-Lie groups of operators on a Hilbert space.

L*-algebras and classical Lie algebras of operators

The starting point of our work was Schue's classification of complex L*-algebras [153], [154]. By definition, a L*-algebra over $\mathbb{K}$ ($\mathbb{K}$ is $\mathbb{R}$ or $\mathbb{C}$) is both an involutive Lie algebra and a $\mathbb{K}$-Hilbert space such that the following holds : if $X \mapsto X^*$ denotes the involution and $\langle\langle | \rangle\rangle$ the scalar product of $\mathfrak{g}$, then $\langle\langle [X,Y] | Z \rangle\rangle = \langle\langle Y | [X^*,Z] \rangle\rangle$ for all $X,Y,Z \in \mathfrak{g}$; and $\mathfrak{g}$ is said to be semi-simple if moreover its derived ideal is dense : $\overline{[\mathfrak{g},\mathfrak{g}]} = \mathfrak{g}$. It is an easy corollary of theorem 3 that any L*-algebra is enlargable; a L*-group is a Banach-Lie group whose Lie algebra has a structure of L*-algebra.

It is relatively easy to show that a semi-simple L*-algebra is the Hilbert direct sum of its simple closed ideals [153]. Schue gave moreover a complete classification of the separable complex L*-algebras. As in finite dimensions, the classification of the separable simple real L*-algebras can then easily be reduced to the following problem : classify the real forms of a small number of explicitely given complex L*-algebras. We solved this last problem in [76]; the results were found independently by Balachandran [15] and Unsain [174], [175].

The use of Hilbert space techniques is crucial for the general theory of L*-algebras. However, the structure of a L*-algebra is unnecessarily restrictive as soon as one considers problems about explicitely given algebras. This is the reason why the whole of Chapter I is a systematic study of Lie algebras of finite rank operators on a (real or) complex Hilbert space $\mathcal{H}$. In Chapter II, we can then consider various Banach-Lie algebras of bounded operators (sections 2-4) and of compact operators (sections 5-6); these include among others all the separable simple L*-algebras.

Derivations, automorphisms, real forms

Those algebras that we call classical Lie algebras of operators are defined by enumeration on pages I.5 and I.46 (algebras of finite rank operators), pages II.10. and II.15 (algebras of bounded operators), and pages II.20 and II.23 (algebras of compact operators). The classification result then reads :

Theorem 4 (= propositions I.12, II.4 and II.13 of the text). Let $\mathfrak{g}$ be a classical complex Lie algebra of finite rank operators (resp. bounded operators, compact operators)on the complex Hilbert space $\mathcal{H}$ and let $\mathfrak{s}$ be a real form of $\mathfrak{g}$. Then $\mathfrak{s}$ is *-isomorphic to one of the classical real Lie algebras listed page I.46 (resp. II.15, II.23).

Theorem 4 is essentially an extension of the analogous finite dimensional result of E. Cartan (see Helgason [84], chap. IX § 4). It follows from the

Theorem 5 (= proposition I.10, II.3 and II.12). Let $\mathfrak{g}$ be as in theorem 4 and let φ be an automorphism of $\mathfrak{g}$ such that $\varphi(X^*) = \varphi(X)^*$ for all $X \in \mathfrak{g}$. Then there exists either a unitary operator V on $\mathcal{H}$ such that $\varphi(X) = VXV^*$ for all $X \in \mathfrak{g}$, or an antiunitary operator $\tilde{V}$ on $\mathcal{H}$ such that $\varphi(X) = -\tilde{V}X^*\tilde{V}^*$ for all $X \in \mathfrak{g}$. In particular, if $\mathfrak{g}$ is a classical complex Banach-Lie algebra of bounded (resp. compact) operators, then any *-automorphism of $\mathfrak{g}$ is isometric and inner (resp. spatial).

The other results in Chapters I and II are:

Any derivation of a classical complex Lie algebra of operators is spatial; in particular, any derivation of a classical complex Banach-Lie algebra of bounded or compact operators is continuous. (Propositions I.2, II.2 and II.9.) This was partially announced in [80].

A conjugation theorem about Cartan subalgebras (proposition I.3 and remark page I.47), and a description of the associated systems of roots (section I.4). This is due, in the L*-context, to Balachandran [12].

A partial classification of the ideals in the classical complex Banach-Lie algebras of bounded operators (section II.2) which is essentially as section 5 of [126].

Most of the propositionsin Chapters I and II fall into three cases corresponding respectively to the general linear groups (type A), the orthogonal complex groups (type B) and the symplectic complex groups (type C). The results for type A can often be proved from theorems on associative algebras (as in Dixmier [46], chap. III §9) and from theorems on the Lie structure of associative rings (as in Herstein [86], and Martindale [117], [118]); but our methods are original, and apply also to types B and C.

Some other considerations (on systems of roots, on classical groups) are more or less known, but it would be difficult to provide a short list of adequate references, either for the material or for the viewpoint.

Cohomology

The Classical Banach-Lie groups of compact operators and the Grassmann manifolds defined by them provide models for the classifying spaces of finite dimensional vector bundles (section III.1). Hence it is natural to study the relationship between the cohomologies of these groups, of their Lie algebras and of their classifying spaces. The cohomology of a Banach-Lie algebra is defined in the standard way (as in Koszul [100]) with the restriction that the cochains are continuous; we consider scalar cohomology only. In this introduction, we consider algebras of type A only; see chapter IV for analogous results about types B and C.

Let $\mathcal{H}$ be an infinite dimensional complex Hilbert space, let $p \in \bar{\mathbb{R}}$ with $1 \leq p \leq \infty$, let $C_p(\mathcal{H})$ be one of the Schatten's minimal ideals (see [151] and the second appendix to Chapter II) and let $\underline{gl}(\mathcal{H}; C_p)$ be the corresponding classical complex Banach-Lie algebra of compact operators.

<u>Theorem 6</u> (= proposition IV.1 and corollaries). Let $J_c^*(\underline{g})$ be the algebra of <u>invariant continuous cochains</u> on the Lie algebra $\underline{g} = \underline{gl}(\mathcal{H}; C_p)$. Then $J_c^*(\underline{g})$ is an <u>exterior algebra</u> generated by a family $(\hat{\jmath}_k)_{k \in N,\ 2k-1 \geq p}$ of primitive cocycles of odd degrees, each of them being unique up to multiplication by a scalar. In particular, if $\underline{g}$ is the Lie algebra defined by the minimal ideal of all compact operators (i.e. if $p = \infty$), then $J_c^*(\underline{g}) \approx \mathbb{C}$.

If $\underline{g}$ was a finite dimensional classical Lie algebra, it is well known that $J^*(\underline{g})$ would be isomorphic (as a ring) to the de Rham cohomology of the classical group defined by $\underline{g}$ (Chevalley-Eilenberg [38]). In infinite dimensions, theorem 6 implies the following

<u>Corollary</u> (= proposition IV.2). Let $\underline{g}$ be as in theorem 6 and let G be the corresponding classical Banach-Lie group. Then $J_c^*(\underline{g})$ is isomorphic to the real cohomology of G is and only if $p = 1$.

The computation of $J_c^*(\underline{g})$ is relatively easy. That of the cohomology algebra $H_c^*(\underline{g})$ seems considerably more difficult; however:

Theorem 7 (= proposition IV.3). Let $\underline{g}$ be as in theorem 6 and let $b_k(p)$ be the dimension of the cohomology space $H_c^k(\underline{g})$. Then :

$$b_1(p) = \begin{cases} 1 & \text{if } p = 1 \\ 0 & \text{otherwise} \end{cases}$$

$$b_2(p) = \begin{cases} 0 & \text{if } p = 1 \text{ or if } p = \infty \\ \infty & \text{otherwise.} \end{cases}$$

If $\underline{g}$ was a finite dimensional classical Lie algebra, it is well known that $J^*(\underline{g})$ would be isomorphic to $H^*(\underline{g})$. In infinite dimensions, theorem 7 implies the following

Corollary. Let $\underline{g}$ be as in theorem 6 and suppose that p is different from 1. Then $J_c^*(\underline{g})$ and $H_c^*(\underline{g})$ are not isomorphic.

Conjecture. When $p = 1$, i.e. when $\underline{g}$ is the Banach-Lie algebra of trace class operators $\underline{gl}(\mathcal{H}; C_1)$, then $J_c^*(\underline{g})$ and $H_c^*(\underline{g})$ are isomorphic.

Finally, let $I_c(G)$ be the algebra of those continuous scalar-valued polynomial functions on $\underline{g} = \underline{gl}(\mathcal{H}; C_p)$ which are invariant by the adjoint action of $G = GL(\mathcal{H}; C_p)$.

Theorem 8 (= proposition IV.4). Let $G = GL(\mathcal{H}; C_p)$. Then $I_c(G)$ is a polynomial algebra generated by a family $(\mathcal{J}_k)_{k \in N, k \geq p}$ of functions of even degrees. In particular, if G is the Fredholm group of $\mathcal{H}$ (i.e. if $p = \infty$), then $I_c(G) \approx \mathbf{C}$.

If G was a finite dimensional classical Lie group, it is well known that I(G) would be isomorphic (as a ring) to the real cohomology of the classifying space B_G. In infinite dimensions, theorem 8 implies the following

Corollary. Let G be as in theorem 8. Then $I_c(G)$ is isomorphic to the real cohomology of the classifying space B_G if and only if $p = 1$.

Theorems 6, 7 and 8 were partially announced in [82] and [83].

Background

As it must be clear after this section 3 of the introduction, our work relies heavily on both the theory of finite dimensional semi-simple Lie algebras over ℂ or ℝ and the theory of associative Banach algebras of operators. We refer for the former essentially to Bourbaki [25], Helgason [84] chap. II and III, Serre [157], and occasionally to Bourbaki [26], Chevalley [36], [37], Jacobson [90], Séminaire Sophus Lie [156]. We refer for the later essentially to Schatten [151], and occasionally to Dieudonné [45] chap. XV, Dixmier [46], Calkin [30], Johnson-Sinclair [92].

0.4.- Homogeneous spaces of the classical Banach-Lie groups, symmetric spaces (Chapter III.)

An analytic Riemannian manifold M is called a Riemannian globally symmetric space if any of its points $m \in M$ is an isolated fixed point of some isometry s_m of M (Helgason [84]). Though this definition looks quite convenient for Hilbert manifolds, little is known about such infinite dimensional spaces in general. However, we seem to be in a good position for further studies because of the following facts:

One of the achievements of the finite dimensional theory of symmetric spaces is the classification of E. Cartan; it makes it possible, among other things, to check conjectures by inspection. Its three main steps are : the classification of complex semi-simple Lie algebras, the classification of the real forms of these and the classification of the symmetric spaces themselves. Suppose such a pattern carries over to the infinite dimensional Hilbert case; then step one is due to Schue and step two is theorem 4 above; so that step three is a natural problem to consider. Short of having solved it, we give in Chapter III the (hopefully complete list of) examples of irreducible Hilbert Riemannian globally symmetric spaces. Emphasis is put on those which correspond to the finite dimensional ones of the compact type; those of non-compact type can be dealt with the same way. The homotopy type (hence the homology) of these spaces is recalled in propositions III.1 and III.2. (Section III.1.)

We then check that various notions related to that of a symmetric space make sense in infinite dimensions : orthogonal symmetric Lie algebra, Hilbert Riemannian symmetric pair, duality. Geometrical properties of the infinite dimensional spaces are then shown, still on examples, to be much the same as in finite dimensions : curvature, geodesics (proposition III.3), geodesic subspaces (proposition III.4). (Section III.2.)

Section III.3 points to an unexpected fact : Let M be an irreducible Hilbert Riemannian globally symmetric space, or for that matter a "stable" irreducible Riemannian globally symmetric space in the same sense as one speaks of "stable classical groups" (because of homotopy equivalences as in propositions II.16 and III.2). Let $P_M(t)$ be the formal power series whose k^{th} coefficient is the kth Betti number of M. Then $P_M(t)$ defines an holomorphic function in the open unit disc of the complex plane. This function is either a polynomial function, or is suprisingly simply related to modular functions.

The last section of Chapter III is made of two independent remarks about particular symmetric spaces. The first recalls the classification of the (simply connected) spaces of constant curvature in the Hilbert case. The second gives an explicit trivialization of the tangent bundle to the unit sphere of a Hilbert space.

0.5.- Acknowledgement

This research has been proposed to me, constantly encouraged and more than once sorted out by J. Eells.

Considerable help has been found in working with or listening to K. D. Elworthy, V. L. Hansen, G. Lusztig, H. Omori, R. Ramer, P.Stefan I. Stewart and many others.

My work has been financially supported by a scholarship of the "Fonds National Suisse pour la Recherche Scientifique" from the Autumn 1968 till the Summer 1971, and by the "Société de Belles-Lettres de Lausanne" during the Autumn 1971.

It is a pleasure for me to thank all of them for their respective contributions, not forgetting the typing of Mrs. J. Lloyd and Mrs. J. Pladdys.

DETAILED TABLE OF CONTENTS

Chapter II : Classical involutive Banach-Lie algebras and groups of bounded and compact operators

0.21

SOME NOTATIONS AND CONVENTIONS

$\mathbb{N}$	set of natural integers : $\{0, 1, 2, \ldots\}$
$\mathbb{N}^*$	set of strictly positive integers : $\{1, 2, \ldots\}$
$\mathbb{Z}$	group of rational integers.
$\mathbb{Z}^*$	set of non zero rational integers.
δ_{ij}	Kronecker symbol.
$\mathbb{R}$	field of real numbers.
$\overline{\mathbb{R}}$	extended real line : $\{-\infty\} \cup \mathbb{R} \cup \{+\infty\}$, together with its usual total order.
$\mathbb{C}$	field of complex numbers.
$\mathbb{Q}$	field of quaternionic numbers.
$\mathbb{K}$	one of the fields $\mathbb{R}$, $\mathbb{C}$, $\mathbb{Q}$.
$\mathbb{K}^*$	set of non zero elements in $\mathbb{K}$.

* * * * *

$E_{\mathbb{K}}$	Banach space over $\mathbb{K}$.
$\mathrm{Lin}(E_{\mathbb{K}})$	space of all (not necessarily bounded) linear maps from $E_{\mathbb{K}}$ into itself; it is an associative algebra over $\mathbb{R}$ if $\mathbb{K}$ is $\mathbb{R}$ or $\mathbb{Q}$ and over $\mathbb{C}$ if $\mathbb{K} = \mathbb{C}$.
$L(E_{\mathbb{K}})$	Banach algebra of all bounded linear maps from $E_{\mathbb{K}}$ into itself, i.e. of all operators on $E_{\mathbb{K}}$.
$C(E_{\mathbb{K}})$	Banach algebra of all compact operators on $E_{\mathbb{K}}$.
$C_o(E_{\mathbb{K}})$	associative algebra of finite rank operators on $E_{\mathbb{K}}$.

* * * * *

$\mathcal{H}_{\mathbb{K}}$	Hilbert space over $\mathbb{K}$, denoted by $\mathcal{H}$ when there is no risk of confusion; the scalar product in $\mathcal{H}$ is denoted by $\langle \mid \rangle$.
$C_p(\mathcal{H}_{\mathbb{K}})$	with $p \in \overline{\mathbb{R}}$, $1 \leqslant p \leqslant \infty$, is one of Schatten's norm ideals of compact operators on $\mathcal{H}_{\mathbb{K}}$.
$C_\infty(\mathcal{H}_{\mathbb{K}})$	means the same as $C(\mathcal{H}_{\mathbb{K}})$.

$C_2(\mathcal{H}_{\mathbb{K}})$ is itself a Hilbert space for the scalar product defined by $\langle\langle X|Y\rangle\rangle = \mathrm{trace}(XY^*)$.

$x\otimes\bar{y}$ with $x,y \in \mathcal{H}_{\mathbb{K}}$ is the operator of rank one defined on $\mathcal{H}_{\mathbb{K}}$ by $z \mapsto \langle z|y\rangle x$.

X^* the adjoint of the operator X on $\mathcal{H}_{\mathbb{K}}$.

$X|E$ the restriction of an operator to a subset E of $\mathcal{H}_{\mathbb{K}}$.

$J_{\mathbb{R}}$ conjugation on $\mathcal{H}_{\mathbb{C}}$ (see appendix to Chapter I).

$J_{\mathbb{Q}}$ anticonjugation on $\mathcal{H}_{\mathbb{C}}$ (id.).

* * * * *

Lie groups are denoted by capital letters as G, SO(k), $Sp(\mathcal{H}_{\mathbb{Q}}; C)$.

Lie algebras are denoted by underlined small letters as $\underline{g}$, $\underline{so}(k)$, $\underline{sp}(\mathcal{H}_{\mathbb{Q}}; C)$.

The connected component of the origin of a group as $O(\mathcal{H}_{\mathbb{R}}; C_2)$ is denoted by $O^+(\mathcal{H}_{\mathbb{R}}; C_2)$.

Classical Lie groups and Lie algebras of finite dimensions are denoted as in Helgason [84], chap. IX §4.

Derivations are usually denoted by Δ, automorphisms by φ. A Cartan subalgebra of a Lie algebra $\underline{g}$ is usually denoted by $\underline{h}$, and $\mathcal{R}$ is the set of non-zero roots of $\underline{g}$ with respect to $\underline{h}$.

* * * * *

[187.525] refers to the item n° 187.525 in the bibliography.

∎ indicates the end or the omission of a proof.

CHAPTER I.

CLASSICAL INVOLUTIVE LIE ALGEBRAS OF FINITE RANK OPERATORS

Let $\mathcal{H}_{\mathbb{K}}$ be a Hilbert space over $\mathbb{K}$, of infinite dimension if not otherwise stated. The associative algebra of finite rank operators on $\mathcal{H}_{\mathbb{K}}$ defines a Lie algebra which will be denoted by $\underline{gl}(\mathcal{H}_{\mathbb{K}}; C_o)$; it is a real Lie algebra if $\mathbb{K}$ is $\mathbb{R}$ or $\mathbb{Q}$ and a complex one if $\mathbb{K} = \mathbb{C}$. The aim of this first chapter is to study various subalgebras of $\underline{gl}(\mathcal{H}_{\mathbb{K}}; C_o)$ and, when $\mathbb{K} = \mathbb{C}$, their real forms. For part of what follows, the Hilbert space structure on the vector space $\mathcal{H}_{\mathbb{K}}$ is not needed; see for example Stewart [168], section 4.4.

The main references for this chapter are Balachandran [12] and de la Harpe [76], which were devoted to the computations of, respectively, sections 3-4 and sections 5-6. However, the point of view here is quite different, as no topology is introduced on the Lie algebras of this chapter.

I.1.- Classical complex Lie algebras of finite rank operators

In this section, the base field is $\mathbb{C}$ if not otherwise stated.

Let $\underline{sl}(\mathcal{H}; C_0)$ be the complex Lie algebra of finite rank operators with zero trace on the complex Hilbert space $\mathcal{H}$. It is the derived ideal of $\underline{gl}(\mathcal{H}; C_0)$, where it is of codimension 1.

Lemma 1. Any finite subset of $\underline{sl}(\mathcal{H}; C_0)$ is contained in a finite dimensional simple subalgebra of $\underline{sl}(\mathcal{H}; C_0)$.

Proof. Let $\{X_1, \ldots, X_n\}$ be a finite subset of $\underline{sl}(\mathcal{H}; C_0)$. For each $j \in \{1, \ldots, n\}$, $\ker X_j$ is finite codimensional and $\operatorname{im} X_j$ is finite dimensional in $\mathcal{H}$. Hence there exists a finite dimensional subspace F in $\mathcal{H}$, of dimension at least two, such that $\ker X_j \subset F^{\perp}$ and $\operatorname{im} X_j \subset F$ for all $j \in \{1, \ldots, n\}$. Let $\underline{sl}(F)$ be identified with the subalgebra of $\underline{sl}(\mathcal{H}; C_0)$ consisting of those operators on $\mathcal{H}$ which map F into itself and $F^{\perp}$ onto zero. Then $\{X_1, \ldots, X_n\}$ is contained in $\underline{sl}(F)$.

The property of the Lie algebra $\underline{sl}(\mathcal{H}; C_0)$ stated in lemma 1 is sometimes called "local simplicity".

Proposition 1A. The Lie algebra $\underline{sl}(\mathcal{H}; C_0)$ is simple.

Proof. Let $\underline{a}$ be a non zero ideal in $\underline{sl}(\mathcal{H}; C_0)$. Suppose that $\underline{a}$ is not trivial, and let $X \in \underline{a}$, $X \neq 0$, $Y \in \underline{sl}(\mathcal{H}; C_0)$, $Y \notin \underline{a}$. According to the lemma, there exists a finite dimensional simple subalgebra of $\underline{sl}(\mathcal{H}; C_0)$, say $\underline{s}$, which contains X and Y. Then $\underline{s} \cap \underline{a}$ is a non trivial ideal of $\underline{s}$, which is absurd. Hence $\underline{a} = \underline{sl}(\mathcal{H}; C_0)$. ■

Let $J_{\mathbb{R}}$ be a conjugation on $\mathcal{H}$. Let $\varphi_{\mathbb{R}}$ be the map

$$\begin{cases} C_0(\mathcal{H}) \longrightarrow C_0(\mathcal{H}) \\ X \longmapsto J_{\mathbb{R}} X^* J_{\mathbb{R}} \end{cases}$$

; $\varphi_{\mathbb{R}}$ is an involutive antiautomorphism of the associative algebra $C_0(\mathcal{H})$. The orthogonal complex Lie algebra corresponding to $\underline{sl}(\mathcal{H}; C_0)$ and given by $\varphi_{\mathbb{R}}$ is by definition the Lie algebra

(1) $\underline{o}(\mathcal{H}, J_{\mathbb{R}}; C_o) = \{X \in \underline{sl}(\mathcal{H}; C_o) \mid \varphi_R(X) = -X\}$. If $\mathcal{H}$ is of finite dimension and if $e = (e_1, \ldots, e_n)$ is a $J_{\mathbb{R}}$-basis of type zero in $\mathcal{H}$ (see appendix), operators in $\underline{o}(\mathcal{H}, J_{\mathbb{R}}; C_o)$ are exactly those whose matrix representation with respect to e is skew-symmetric; in this case, the orthogonal complex Lie algebra is denoted usually by $\underline{so}(n, \mathbf{C})$ [84].

Let J_Q be an anticonjugation in $\mathcal{H}$. Let φ_Q be the map $X \mapsto -J_Q X^* J_Q$; φ_Q is again an involutive antiautomorphism of the associative algebra $C_o(\mathcal{H})$. The symplectic complex Lie algebra corresponding to $\underline{sl}(\mathcal{H}; C_o)$ and given by φ_Q is by definition the Lie algebra

(2) $\underline{sp}(\mathcal{H}, J_Q; C_o) = \{X \in \underline{sl}(\mathcal{H}; C_o) \mid \varphi_Q(X) = -X\}$. In the case $\mathcal{H}$ is finite (hence even) dimensional, it is easy to check that $\underline{sp}(\mathcal{H}, J_Q; C_o)$ is the algebra usually denoted by $\underline{sp}(n, \mathbf{C})$.

<u>Proposition 1BC</u>. The Lie algebras $\underline{o}(\mathcal{H}, J_{\mathbb{R}}; C_o)$ and $\underline{sp}(\mathcal{H}, J_Q; C_o)$ are simple.

<u>Proof</u> : as for proposition 1A; in the proof of lemma 1, the space F can now be chosen invariant by J_R or J_Q . ■

<u>Remarks</u>.

i) Proposition 1 is a particular case of much more general results : it follows from the simplicity of the associative algebra $C_o(\mathcal{H})$, and from known facts about the Lie structure of simple rings; see Herstein [86], principally theorems 4, 8 and 10.

ii) Analogues of proposition 1, for groups of the kind

$$SL(\mathcal{H}; C_o) = \left\{ X \in GL(\mathcal{H}) \;\middle|\; \begin{array}{c} X = \text{identity} + \text{finite rank operator} \\ \text{and} \quad \det(X) = 1 \end{array} \right\}$$

are known in various cases to algebraists; see e.g. [39], [143] and [144].

Several of the notions used later on in this work cannot be defined for arbitrary complex Lie algebras, though they are very useful in numerous examples. The two following definitions will allow us to restrict the class of algebras under consideration.

Definition 1. Let $\mathfrak{g}$ be a Lie algebra over R or C. An involution in $\mathfrak{g}$ is a semi-linear map $X \mapsto X^*$ from $\mathfrak{g}$ to $\mathfrak{g}$ such that $(X^*)^* = X$ for all $X \in \mathfrak{g}$ and such that $[X,Y]^* = [Y^*,X^*]$ for all $X,Y \in \mathfrak{g}$. A Lie algebra furnished with an involution is an involutive Lie algebra. A self-adjoint subset of an involutive algebra is a subset globally invariant by the involution. A normal element in an involutive algebra is an element X such that $[X,X^*] = 0$.

Let $\mathfrak{g}$ be a complex Lie algebra; then $\mathfrak{g}$ admits an involution if and only if $\mathfrak{g}$ has a real form (see e.g. Helgason [84] chap. III §6). There exist finite dimensional complex Lie algebras which do not admit any involution (see Bourbaki [25] §5 exercise 8c).

Definition 2. Let V be a vector space over $\mathbb{R}$ or $\mathbb{C}$ (possibly infinite dimensional). Let X be an endomorphism of V and let $\{X\}^+$ be the associative subalgebra of End(V) generated by X. Then X is said to be semi-simple if $\{X\}^+$ does not contain any non zero nilpotent element.

Let $\mathfrak{g}$ be a Lie algebra over $\mathbb{R}$ or $\mathbb{C}$. An element X in $\mathfrak{g}$ is said to be semi-simple if ad(X) is semi-simple as endomorphism of $\mathfrak{g}$. A c-involution in $\mathfrak{g}$ is an involution in $\mathfrak{g}$ such that any normal element with respect to it is semi-simple. A Lie algebra furnished with a c-involution is called a c-involutive Lie algebra.

The definition of a semi-simple element coincides with the standard one when V or $\underline{g}$ is finite dimensional.

Example. Let $\underline{g}$ be a semi-simple complex Lie algebra of finite dimension and let τ be an involution in $\underline{g}$. If $\{X \in \underline{g} \mid \tau(X) = -X\}$ is a compact real form of $\underline{g}$, then τ is a c-involution.

Indeed, define $\langle\langle\,|\,\rangle\rangle \begin{cases} \underline{g} \times \underline{g} \longrightarrow \mathbb{C} \\ (X,Y) \mapsto (B(X) \mid \tau(Y)) \end{cases}$, where B is the Killing form of $\underline{g}$. Then $\langle\langle\,|\,\rangle\rangle$ is a scalar product on $\underline{g}$, and if X is an element of $\underline{g}$, the adjoint of $ad(X)$ (in the Hilbert space sense) is precisely $ad(\tau(X))$. Hence X is normal if and only if $ad(X)$ is a normal operator on the Hilbert space $\underline{g}$, that is if and only if $ad(X)$ is semi-simple.

Example. Let $\underline{g}$ be a semi-simple real Lie algebra of finite dimension and let τ be an involution in $\underline{g}$; let $k = \{X \in \underline{g} \mid \tau(X) = -X\}$ and $\underline{p} = \{X \in \underline{g} \mid \tau(X) = X\}$. If $\underline{k} \oplus \underline{p}$ is a Cartan decomposition of $\underline{g}$, then τ is a c-involution on $\underline{g}$.

The proof mimics that of the previous example.

Example. Let $\underline{g}$ be one of the Lie algebras $\underline{sl}(\mathcal{H}; C_0)$, $\underline{o}(\mathcal{H}, J_{\mathbb{R}}; C_0)$ and $\underline{sp}(\mathcal{H}, J_{\mathbb{Q}}; C_0)$, where $\mathcal{H}$ is some complex Hilbert space. Then $\underline{g}$ has a natural involution given by $X \mapsto X^*$, where X^* is the adjoint of the operator X. This involution is clearly a c-involution in $\underline{g}$.

Definition 3. A classical complex Lie algebra of finite rank operators is one of the c-involutive Lie algebras $\underline{sl}(\mathcal{H}; C_0)$, $\underline{o}(\mathcal{H}, J_{\mathbb{R}}; C_0)$ and $\underline{sp}(\mathcal{H}, J_{\mathbb{Q}}; C_0)$, where $\mathcal{H}$ is a complex Hilbert space of arbitrary dimension.

The algebras of definition 3 enlarge somehow the list of the finite dimensional classical complex Lie algebras, and the following sections will show how they share certain properties of these.

1.2. - Derivations

In this section, $\mathcal{H}$ is a complex Hilbert space (of infinite dimension). An earlier stage of the results below has been summed up in [80].

Lemma 2. Let $\underline{g}$ be a finite dimensional complex Lie algebra and let $\underline{s}$ be a semi-simple subalgebra of $\underline{g}$. Let $\Delta : \underline{s} \to \underline{g}$ be a derivation. Then there exists $D \in \underline{g}$ such that $\Delta(X) = [D,X]$ for all $X \in \underline{s}$.

N.B. : a derivation from $\underline{s}$ into $\underline{g}$ is, in more sophisticated terms, a skew derivation of type (j,j), where j is the inclusion of $\underline{s}$ in $\underline{g}$; see Chevalley [37] chap. I §3.

Proof. Consider $\underline{g}$ as the $\underline{s}$-module defined by $X_{\underline{g}}Z = [X,Z]$ for all $X \in \underline{s}$, $Z \in \underline{g}$. A 1-dimensional $\underline{g}$-cochain on $\underline{s}$ is a linear map $\omega : \underline{s} \to \underline{g}$. It is a cocycle if and only if it is a derivation and it is a coboundary if and only if it is an "inner" derivation, namely if there exists $D \in \underline{g}$ such that $\Delta(X) = [D,X]$ for all $X \in \underline{s}$. Lemma 2 is then a standard and trivial consequence of the first Whitehead lemma; see for example Bourbaki [25] §6 exercice 1a or Jacobson [90] chap. III Lemma 3. ∎

Lemma 3. A. - Let Δ be a derivation of $\underline{sl}(\mathcal{H} ; C_o)$. For every finite dimensional subspace E of $\mathcal{H}$ of dimension at least 2, there exists an operator $Y_E \in \underline{sl}(\mathcal{H} ; C_o)$ such that

(3) $\Delta(X) = [Y_E,X]$ for all $X \in \underline{sl}(E)$.

Moreover, the restriction of Y_E to E is uniquely defined by these properties, up to the addition of a scalar multiple of the identity.

B. - Let Δ be a derivation of $\underline{o}(\mathcal{H}, J_{\mathbb{R}}; C_o)$. For every finite dimensional $J_{\mathbb{R}}$-invariant subspace E of $\mathcal{H}$ of dimension at least 5, there exists an operator $Y_E \in \underline{o}(\mathcal{H}, J_{\mathbb{R}}; C_o)$ such that

(4) $\Delta(X) = [Y_E, X]$ for all $X \in \underline{o}(E, J_{\mathbb{R}})$.

Moreover, the restriction of Y_E to E is uniquely defined by these properties.

C. - Let Δ be a derivation of $\underline{sp}(\mathcal{H}, J_{\mathbb{Q}}; C_o)$. For every finite dimensional $J_{\mathbb{Q}}$-invariant subspace E of $\mathcal{H}$ of dimension at least 2, there exists an operator

$Y_E \in \underline{sp}(\mathcal{H}, J_{\mathbb{Q}}; C_o)$ such that

(5) $\Delta(X) = [Y_E, X]$ for all $X \in \underline{sp}(E, J_{\mathbb{Q}})$.

Moreover, the restriction of Y_E to E is uniquely defined by these properties.

Proof (case A). As the image of $\underline{sl}(E)$ by Δ is finite dimensional, there exists a finite dimensional subspace F in $\mathcal{H}$, orthogonal to E, and such that $\Delta(\underline{sl}(E)) \subset \underline{sl}(E \oplus F)$ (see lemma 1). By lemma 2 and as $\underline{sl}(E)$ is simple as soon as $\dim E \geqslant 2$, there exists $Y_E \in \underline{sl}(E \oplus F)$ such that (3) holds. Let now Y_E and Y'_E be two operators in $\underline{sl}(\mathcal{H}; C_o)$ such that (3) holds for both of them. Let the matrix of $Y_E - Y'_E$ relatively to the decomposition $E \oplus E^{\perp}$ of $\mathcal{H}$ be denoted by

$\begin{pmatrix} a_E & b_E \\ c_E & d_E \end{pmatrix}$. As $\left[Y_E - Y'_E, X\right] = 0$ for all $X \in \underline{sl}(E)$, it follows

that $\begin{cases} [a_E, X] = 0 \\ -Xb_E = 0 \\ c_E X = 0 \end{cases}$ for all $X \in \underline{sl}(E)$. Hence a_E is a scalar

multiple of the identity of E, and b_E and c_E vanish. This ends the proof of case A.

The cases B and C are proved the same way; note that $\underline{o}(E, J_{\mathbb{R}})$ is simple as soon as $\dim E \geqslant 5$ and that $\underline{sp}(E, J_{\mathbb{Q}})$ is simple as soon as $\dim E \geqslant 2$. ∎

The associative algebra of all linear maps (not necessarily bounded) of $\mathcal{H}$ into itself is denoted by $\mathrm{Lin}(\mathcal{H})$.

Lemma 4. Let $D \in \mathrm{Lin}(\mathcal{H})$; suppose that one of the following holds:

A. - D commutes with all operators in $\underline{sl}(\mathcal{H}\,;\,C_o)$.

B. - D commutes with all operators in $\underline{o}(\mathcal{H}\,,J_{\mathbb{R}};\,C_o)$.

C. - D commutes with all operators in $\underline{sp}(\mathcal{H}\,,J_{\mathbb{Q}};\,C_o)$.

Then D is a multiple of the identity of $\mathcal{H}$.

Proof: standard. ∎

Proposition 2.

A. - Let Δ be a derivation of $\underline{sl}(\mathcal{H}\,;\,C_o)$. Then there exists $D \in \mathrm{Lin}(\mathcal{H})$ such that $\Delta(X) = [D,X]$ for all $X \in \underline{sl}(\mathcal{H}\,;\,C_o)$. Moreover, D is uniquely defined by these properties, up to addition of a scalar multiple of $id_{\mathcal{H}}$.

B. - Let Δ be a derivation of $\underline{o}(\mathcal{H}\,,J_{\mathbb{R}};\,C_o)$. Then there exists $D \in \mathrm{Lin}(\mathcal{H})$ such that $\langle Dx \mid y\rangle = -\langle x \mid J_{\mathbb{R}}DJ_{\mathbb{R}}y\rangle$ for all $x,y \in \mathcal{H}$ and such that $\Delta(X) = [D,X]$ for all $X \in \underline{o}(\mathcal{H}\,,J_{\mathbb{R}};\,C_o)$. Moreover, D is uniquely defined by these properties.

C. - Let Δ be a derivation of $\underline{sp}(\mathcal{H}\,,J_{\mathbb{Q}};\,C_o)$. Then there exists $D \in \mathrm{Lin}(\mathcal{H})$ such that $\langle Dx \mid y\rangle = +\langle x \mid J_{\mathbb{Q}}DJ_{\mathbb{Q}}y\rangle$ for all $x,y \in \mathcal{H}$ and such that $\Delta(X) = [D,X]$ for all $X \in \underline{sp}(\mathcal{H}\,,J_{\mathbb{Q}};\,C_o)$. Moreover, D is uniquely defined by these properties.

Proof (case A). Let F be an arbitrary 2-dimensional subspace of $\mathcal{H}$ and let f be a non zero vector in F. For any vector $x \in \mathcal{H}$, let E be the space span by F and x and let Y_E be as in lemma 3A, and such that $Y_E f \perp f$; such an operator Y_E is uniquely defined (the condition on f is only to get rid of the "up to addition of a multiple of the identity"-uniqueness). Put then $y = Y_E x$. The map D from $\mathcal{H}$ to $\mathcal{H}$ which sends x to y is well defined and fulfills the required conditions. The unicity part of proposition 2 follows trivially from lemma 4. ∎

Let $\underline{gl}(\mathcal{H}\ ;\ \mathrm{Lin})$ be the Lie algebra of all linear maps from $\mathcal{H}$ into itself, endowed with the evident product; proposition 2 can then be restated as follows : the Lie algebra of the derivations of $\underline{sl}(\mathcal{H}\ ;\ C_o)$ is isomorphic to the quotient of $\underline{gl}(\mathcal{H}\ ;\ \mathrm{Lin})$ by its center $\mathbb{C}\mathrm{id}_{\mathcal{H}}$.

Similarly, the Lie algebra of the derivations of $\underline{o}(\mathcal{H}, J_{\mathbb{R}};\ C_o)$ is isomorphic to : $\underline{o}(\mathcal{H}, J_{\mathbb{R}};\ \mathrm{Lin}) =$
$= \{D \in \underline{gl}(\mathcal{H}\ ;\ \mathrm{Lin})\ |\ \langle Dx \mid y\rangle = -\langle x \mid J_{\mathbb{R}}DJ_{\mathbb{R}}y\rangle$ for all $x,y \in \mathcal{H}\}$.
And that of $\underline{sp}(\mathcal{H}, J_{\mathbb{Q}};\ C_o)$ to : $\underline{sp}(\mathcal{H}, J_{\mathbb{Q}};\ \mathrm{Lin}) =$
$= \{D \in \underline{gl}(\mathcal{H}\ ;\ \mathrm{Lin})\ |\ \langle Dx \mid y\rangle = +\langle x \mid J_{\mathbb{Q}}DJ_{\mathbb{Q}}y\rangle$ for all $x,y \in \mathcal{H}\}$.

Remarks.

i) Proposition 2A is still true if $\underline{sl}(\mathcal{H}\ ;\ C_o)$ is replaced by $\underline{gl}(\mathcal{H}\ ;\ C_o)$. In other words, if Δ is a derivation of $\underline{gl}(\mathcal{H}\ ;\ C_o)$, then Δ is also a derivation of the associative simple algebra $C_o(\mathcal{H})$.

ii) Proposition 2A can alternatively be deduced from general results, due to Martindale [117], about Lie derivations of certain associative rings (see his theorem 2 with $R = C_o(\mathcal{H})$ and $\bar{R} = \mathrm{Lin}(\mathcal{H})$).

iii) The people who live in $\underline{o}(\mathcal{H}, J_{\mathbb{R}};\ \mathrm{Lin})$ and $\underline{sp}(\mathcal{H}, J_{\mathbb{Q}};\ \mathrm{Lin})$ are in fact continuous (because they have adjoints; [45], 12. 16. 7). It is not necessarily the case for those in $\underline{gl}(\mathcal{H}\ ;\ \mathrm{Lin})$.

I.3. - Cartan subalgebras

In order to make sense, the definition of Cartan subalgebra given below requires a definition of semi-simplicity for infinite dimensional Lie algebra. (According to tradition, the word "semi-simplicity" is confusingly applied to both Lie algebras and elements inside a Lie algebra.)

For this chapter, a Lie algebra is defined to be semi-simple if it has no non trivial abelian ideals. A critic of this definition will be given in an appendix to chapter II. But certainly, any other reasonable definition of semi-simplicity will imply the absence of non trivial abelian ideals, so that the matter of this chapter will not have to be changed in any respect whatsoever.

Definition 4. Let $\underline{g}$ be a semi-simple c-involutive Lie algebra over $\mathbb{K}$. A Cartan subalgebra of $\underline{g}$ is a subalgebra $\underline{h}$ of $\underline{g}$ which is maximal among the abelian self-adjoint subalgebras of $\underline{g}$.

In particular, a Cartan subalgebra of $\underline{g}$ is maximal among the abelian subalgebras of $\underline{g}$, and all its elements are semi-simple. When $\underline{g}$ is finite dimensional, it follows that Cartan subalgebras of $\underline{g}$ according to definition 4 are Cartan subalgebras of $\underline{g}$ in the usual sense.

Conversely, let $\underline{g}$ be a finite dimensional semi-simple complex Lie algebra and let $\underline{h}$ be a Cartan subalgebra of $\underline{g}$ in the usual sense. Then there exists a c-involution in $\underline{g}$ for which $\underline{h}$ is invariant ([156], exposé 11, théorèmes 2 and 3).

Let $\underline{g}$ be a finite dimensional semi-simple real Lie algebra and let $\underline{h}$ be a Cartan subalgebra of $\underline{g}$ in the usual sense. If $\underline{h}$ is standard (Kostant [99]), there is evidently a c-involution in $\underline{g}$ for which $\underline{h}$ is invariant. (Is this still true when $\underline{h}$ is not assumed to be standard?)

The purpose of the present section is to determine the Cartan subalgebras of the classical complex Lie algebras of finite rank operators in a (complex) Hilbert space $\mathcal{H}$. The analysis follows an argument devised by Balachandran [12] for the study of certain L^*-algebras. For simplicity in the notations, $\mathcal{H}$ is supposed to be <u>infinite dimensional and separable</u> in the rest of this section.

<u>Proposition 3A</u>. Let $\underline{h}$ be a Cartan subalgebra in $\underline{sl}(\mathcal{H}; C_o)$. Then there exists an orthonormal **basis** $e = (e_n)_{n \in N}$ of $\mathcal{H}$ such that $\underline{h}$ consists of those operators in $\underline{sl}(\mathcal{H}; C_o)$ which are diagonal with respect to e. In particular, two Cartan subalgebras of $\underline{sl}(\mathcal{H}; C_o)$ are conjugated by an element of the full unitary group $U(\mathcal{H})$.

<u>Proof</u>: immediate via the spectral theorem. ∎

A basis such as e in proposition 4A is said to be <u>compatible</u> with $\underline{h}$.

<u>Proposition 3B</u>. Let $\underline{h}$ be a Cartan subalgebra in $\underline{g} = \underline{o}(\mathcal{H}, J_{\mathbb{R}}; C_o)$. Then :

Either there exists a $J_{\mathbb{R}}$-basis of type 1 in $\mathcal{H}$ (see definition in the appendix), say $f = (f_n)_{n \in Z}$, such that $\underline{h}$ consists of those operators in $\underline{g}$ which are diagonal with respect to f.

Or there exists a $J_{\mathbb{R}}$-basis of type 2 in $\mathcal{H}$, say $g = (g_n)_{n \in Z^*}$, such that $\underline{h}$ consists of those operators in $\underline{g}$ which are diagonal with respect to g.

The two cases exclude each other.

In particular, there are two conjugacy classes of Cartan

subalgebras in $\mathfrak{g}$ under the action of the group
$O(\mathcal{H}_{\mathbb{R}}) = \{X \in U(\mathcal{H}) \mid XJ_{\mathbb{R}} = J_{\mathbb{R}}X\}$.

<u>Lemma 5B</u>. Let $\mathfrak{g}$ be as in proposition 3B and let x be a vector of norm 1 in $\mathcal{H}$ such that $y = J_{\mathbb{R}}x$ is orthogonal to x. Then the self-adjoint operator $T = x \otimes \bar{x} - y \otimes \bar{y}$ belongs to $\mathfrak{g}$. Furthermore, if $e = (e_n)_{n \in N}$ is any orthonormal basis of $\mathcal{H}$ containing x, then T commutes with all the operators in $\mathfrak{g}$ which are diagonal with respect to e.

<u>Proof of lemma 5B</u>.

Let $z \in \mathcal{H}$, z orthogonal to both x and y; then $J_{\mathbb{R}}z$ is orthogonal to both x and y and $(TJ_{\mathbb{R}} + J_{\mathbb{R}}T^*)z = 0$. Trivially, $(TJ_{\mathbb{R}} + J_{\mathbb{R}}T^*)x = 0$ and $(TJ_{\mathbb{R}} + J_{\mathbb{R}}T^*)y = 0$. Hence $T \in \mathfrak{g}$.

Let now e be as in the lemma, with $e_o = x$, and let X be an operator in $\mathfrak{g}$ which is diagonal with respect to e. Suppose first that X is self-adjoint and let $(\xi_n)_{n \in N}$ be the sequence of real numbers such that $Xe_n = \xi_n e_n$ for all $n \in N$. Now

$XJ_{\mathbb{R}}x = X \sum \langle J_{\mathbb{R}}x \mid e_n \rangle e_n = \sum \xi_n \langle y \mid e_n \rangle e_n$ and

$J_{\mathbb{R}}Xx = \xi_o J_{\mathbb{R}}x = \xi_o \sum \langle y \mid e_n \rangle e_n$; hence $\xi_n = -\xi_o$ whenever $\langle y \mid e_n \rangle \neq 0$.

Let $\mathcal{H}_I$ [resp. $\mathcal{H}_{II}$] be the closed span of those e_n for which $\langle y \mid e_n \rangle \neq 0$ [resp. $\langle y \mid e_n \rangle = 0$]. As $\mathcal{H}_I$ is invariant by T and as $X|\mathcal{H}_I$ is a scalar operator, $X|\mathcal{H}_I$ and $T|\mathcal{H}_I$ commute. If $z = \sum z_n e_n \in \mathcal{H}_{II}$, then $XTz = Xz_o e_o = \xi_o z_o e_o$ and

$TXz = \sum \xi_n z_n T(e_n) = \xi_o z_o e_o$, so that $X|\mathcal{H}_{II}$ and $T|\mathcal{H}_{II}$ commute. **Hence** X commutes with T.

The general case follows clearly from the case where X is self-adjoint. ∎

Proof of proposition 3B.

Let $\underline{h}$ be a Cartan subalgebra in $\underline{g}$. From the spectral theorem, it follows that there exists an orthonormal basis $e = (e_n)_{n \in N}$ in $\mathcal{H}$ such that the operators of $\underline{h}$ are diagonal with respect to e. Let $e_I = \{e_n \in e \mid Xe_n = 0 \text{ for all } X \in \underline{h}\}$ and let e_{II} be the set-complement of e_I in e; let $\mathcal{H}_I$ [resp. $\mathcal{H}_{II}$] be the closed span of e_I [resp. e_{II}].

Step one : $J_{\mathbb{R}} \mathcal{H}_I = \mathcal{H}_I$ and $J_{\mathbb{R}} \mathcal{H}_{II} = \mathcal{H}_{II}$.

Let e_k be a vector in e_I and let X be a self-adjoint operator in $\underline{h}$ represented by its diagonal matrix $(\xi_n)_{n \in N}$. Then $0 = J_{\mathbb{R}} X e_k = - \sum \langle J_{\mathbb{R}} e_k \mid e_n \rangle \xi_n e_n$, so that $\xi_n = 0$ if $\langle J_{\mathbb{R}} e_k \mid e_n \rangle \neq 0$; hence $J_{\mathbb{R}} e_k = \sum_n \langle J_{\mathbb{R}} e_k \mid e_n \rangle e_n$ where the last sum is only over those n such that $\xi_n = 0$ for all self-adjoint $X \in \underline{h}$. Hence $J_{\mathbb{R}} e_k \in \mathcal{H}_I$, that is $J_{\mathbb{R}} \mathcal{H}_I \subset \mathcal{H}_I$.

Let now e_ℓ be a vector in e_{II} and let X be a self-adjoint operator of $\underline{h}$ represented by a diagonal matrix $(\xi_n)_{n \in N}$ with $\xi_\ell \neq 0$. As in the proof of lemma 5B, the equality $XJ_{\mathbb{R}} e = - J_{\mathbb{R}} Xe$ implies that $\xi_n = - \xi_\ell$ whenever $\langle J_{\mathbb{R}} e_\ell \mid e_n \rangle \neq 0$. Hence $J_{\mathbb{R}} e_\ell \in \mathcal{H}_{II}$, that is $J_{\mathbb{R}} \mathcal{H}_{II} \subset \mathcal{H}_{II}$.

As $J_{\mathbb{R}}^2 = id_{\mathcal{H}}$, the claim follows. As by product of the proof, one obtains that $J_{\mathbb{R}} e_\ell \perp e_\ell$ for all $e_\ell \in e_{II}$.

Step two : $\dim \mathcal{H}_I \leqslant 1$.

Suppose that $\dim \mathcal{H}_I \geqslant 2$; then there exists a non zero vector $x \in \mathcal{H}_I$ such that $y = J_{\mathbb{R}} x$ is orthogonal to x . The operator $y \otimes \bar{x} - x \otimes \bar{y}$ is a skew-adjoint element of $\underline{g}$ which does not belong to $\underline{h}$ and which commutes with every element of $\underline{h}$. As this is absurd by definition of $\underline{h}$, the claim follows.

Step three : the case $\dim \mathcal{H}_I = 1$.

Assume the base has been re-indexed such that $e = (f_n)_{n \in Z}$ with $\mathcal{H}_I = \mathbb{C} f_o$, and suppose moreover that $J_{\mathbb{R}} f_o = f_o$. Let $n \in Z$ and let $T = f_n \otimes \overline{f_n} - J_{\mathbb{R}} f_n \otimes \overline{(J_{\mathbb{R}} f_n)}$. Suppose first that $T \neq 0$;

then $n \neq 0$, $J_{\mathbb{R}}f_n$ is orthogonal to f_n, and lemma 5B applies. By maximality of Cartan subalgebras, $T \in \underline{h}$, hence $J_{\mathbb{R}}f_n \otimes \overline{(J_{\mathbb{R}}f_n)}$ is diagonal with respect to e, which implies that $J_{\mathbb{R}}f_n$ is parallel to f_m for some $m \in Z^*$, $m \neq n$. Suppose now that $T = 0$; then obviously $n = 0$. Modulo a second re-indexing of the vectors in e and multiplication of each of them by ad hoc constants, $e = (f_n)_{n \in Z}$ can be made a $J_{\mathbb{R}}$-basis of type I.

Step four: the case $\dim \mathcal{H}_I = 0$. The same argument shows how to obtain from e a $J_{\mathbb{R}}$-basis of type II.

Finally, suppose there exists a Cartan subalgebra $\underline{h}$ in $\underline{g}$, a $J_{\mathbb{R}}$-basis $f = (f_n)_{n \in Z}$ of type one and a $J_{\mathbb{R}}$-basis $g = (g_n)_{n \in Z^*}$ of type two in $\mathcal{H}$, such that the elements of $\underline{h}$ are diagonal with respect to both e and f. Choose $g_k \in g$ such that $\langle f_o \mid g_k \rangle \neq 0$ and let $T = g_k \otimes \overline{g_k} - g_{-k} \otimes \overline{g_{-k}}$. Then $0 = Tf_o = \sum \langle f_o \mid g_n \rangle T(g_n) = \langle f_o \mid g_k \rangle g_k - \langle f_o \mid g_{-k} \rangle g_{-k}$, which is absurd. Proposition 3B follows. ■

Definition 5. A Cartan subalgebra $\underline{h}$ in $\underline{g} = \underline{o}(\mathcal{H}, J_{\mathbb{R}}; C_o)$ is said to be of type one [resp. of type two] if it consists of all operators of $\underline{g}$ which are diagonal with respect to some $J_{\mathbb{R}}$-basis of type one [resp. of type two] in $\mathcal{H}$. Such a basis is said to be compatible with $\underline{h}$.

Proposition 3B implies that two Cartan subalgebras of $\underline{g}$ are of the same type if and only if they are conjugated by an element of $U(\mathcal{H})$, if and only if they are conjugated by an element of $O(\mathcal{H}_{\mathbb{R}})$.

Proposition 3C. Let $\underline{h}$ be a Cartan subalgebra in $g = \underline{sp}(\mathcal{H}, J_{\mathbb{Q}}; C_o)$. Then there exists a $J_{\mathbb{Q}}$-basis in $\mathcal{H}$, say $e = (e_n)_{n \in Z^*}$, such that $\underline{h}$ consists of those operators in $\underline{g}$ which are diagonal with respect to e. In particular, two Cartan subalgebras of $\underline{g}$ are conjugated by an element of the group $Sp(\mathcal{H}_Q) = \{X \in U(\mathcal{H}) \mid XJ_Q = J_Q X\}$.

Lemma 5C. Let $\underline{g}$ be as in proposition 3C, let x be a vector of norm 1 in $\mathcal{H}$ and let $y = J_Q x$ (y is automatically orthogonal to x). Then the self-adjoint operator $T = x \otimes \bar{x} - y \otimes \bar{y}$ belongs to $\underline{g}$. Furthermore, if $e = (e_n)_{n \in N}$ is any orthonormal basis of $\mathcal{H}$ containing x, then T commutes with all the operators in $\underline{g}$ which are diagonal with respect to e.

Proofs of lemma 5C and of proposition 3C: as for lemma 5B and proposition 3B. ■

Again, a basis such as e in proposition 4C is said to be compatible with $\underline{h}$.

Corollary to proposition 3 : Let $\underline{h}$ be a Cartan subalgebra of a classical complex Lie algebra of finite rank operators. Then $\underline{h}$ is equal to its normalizer.

Proof : immediate. ■

Remark. Let $\underline{g}$ be one of the infinite dimensional Lie algebras $\underline{sl}(\mathcal{H}; C_o)$, $\underline{o}(\mathcal{H}, J_R; C_o)$ and $\underline{sp}(\mathcal{H}, J_Q; C_o)$. Let $\underline{h}$ be a Cartan subalgebra of $\underline{g}$ and let H be an element of $\underline{h}$. It follows trivially from proposition 3 that the normalizer $N(H) = \{X \in \underline{g} \mid [H,X] = 0\}$ is always strictly larger than $\underline{h}$. In particular, $\underline{h}$ contains no regular elements, unlike to what happens either in the finite dimensional case or in the separable L^*-case [9].

I.4. - Roots

Let $\underline{g}$ be a semi-simple c-involutive complex Lie algebra and let $\underline{h}$ be a Cartan subalgebra of $\underline{g}$. For any linear functional $\alpha : \underline{h} \to \mathbb{C}$, write $\underline{g}_\alpha = \{X \in \underline{g} \mid [H,X] = \alpha(H)X \text{ for all } H \in \underline{h}\}$. Then α is called a root of $\underline{g}$ with respect to $\underline{h}$, or simply a root, if $\underline{g}_\alpha \neq 0$; if α is a root, $\underline{g}_\alpha$ is its root space and any non zero vector in $\underline{g}_\alpha$ is a root vector of α. The zero functional is clearly a root, and $\underline{g}_0 = \underline{h}$. The set of nonzero roots of $\underline{g}$ with respect to $\underline{h}$ will be denoted by $\mathcal{R}$.

Lemma 6. Let $\underline{g}$, $\underline{h}$ and $\mathcal{R}$ be as above. Let $\mathcal{R}' = (\alpha_\iota)_{\iota \in I}$ be a subset of $\mathcal{R}$ and, for each $\iota \in I$, let X_ι be a root vector of α_ι. Suppose that the following conditions are satisfied :

i) $\alpha_\iota(H^*) = \overline{\alpha_\iota(H)}$ for all $H \in \underline{h}$, for all $\iota \in I$;

ii) there exists a subspace $\underline{m} \subset \underline{g}$ such that $\underline{g} = \underline{h} \oplus \underline{m}$, such that $[\underline{h},\underline{m}] \subset \underline{m}$ and such that $X_\iota \in \underline{m}$ for all $\iota \in I$;

iii) there exists a sesquilinear form $\langle\langle\,|\,\rangle\rangle : \underline{m} \times \underline{m} \to \mathbb{C}$ such that $\langle\langle [H,X] \mid Y\rangle\rangle = \langle\langle X \mid [H^*,Y]\rangle\rangle$ for all $H \in \underline{h}$ and $X,Y \in \underline{m}$;

iv) the set $(X_\iota)_{\iota \in I}$ is total in $\underline{m}$ with respect to $\langle\langle\,|\,\rangle\rangle$, that is $X \in \underline{m}$ and $\langle\langle X \mid X_\iota\rangle\rangle = 0$ for all $\iota \in I$ implies $X = 0$.

Then $\mathcal{R}' = \mathcal{R}$.

Proof. Let β be a non zero root of $\underline{g}$ with respect to $\underline{h}$. By definition of a root, there exists $X \in \underline{g} - \{0\}$ such that $[H,X] = \beta(h)X$ for all $H \in \underline{h}$. As $\underline{g} = \underline{h} \oplus \underline{m}$ and as $[\underline{h},\underline{m}] \subset \underline{m}$, X lies in $\underline{m}$. Then, for all $\iota \in I$ and for all $H \in \underline{h}$:

$$\beta(H) \langle\langle X \mid X_\iota\rangle\rangle = \langle\langle [H,X] \mid X_\iota\rangle\rangle = \langle\langle X \mid [H^*,X_\iota]\rangle\rangle = \alpha_\iota(H) \langle\langle X \mid X_\iota\rangle\rangle.$$

As $X \neq 0$, there exists $\jmath \in I$ such that $\langle\langle X \mid X_\jmath\rangle\rangle \neq 0$. Hence

$\beta(H) = \alpha_j(H)$ for all $H \in \underline{h}$, that is $\beta = \alpha_j$, and $\beta \in \mathcal{R}'$. ■

For simplicity in the notations again, $\mathcal{H}$ is supposed to be an infinite dimensional separable complex Hilbert space in the rest of this section. When $\mathcal{H}$ is furnished with an orthonormal basis $e = (e_n)_{n \in N}$, and if i and j are two natural integers, $E_{i,j}$ is the operator on $\mathcal{H}$ whose matrix representation with respect to e is given by $(E_{i,j})_{m,n} = \delta_{i,m}\,\delta_{j,n}$ for all $m,n \in N$.

Proposition 4A. Let $\underline{g} = \underline{sl}(\mathcal{H}; C_o)$, let $\underline{h}$ be a Cartan subalgebra of $\underline{g}$ and let $e = (e_n)_{n \in N}$ be a basis of $\mathcal{H}$ compatible with $\underline{h}$. For each $i \in N$, let $\lambda_i : \underline{h} \longrightarrow \mathbf{C}$ be defined by $\lambda_i(H) = \mathrm{trace}(HE_{i,i})$. Then $\mathcal{R} = \{\lambda_i - \lambda_j \in \underline{h}^{dual} \mid i,j \in N,\ i \neq j\}$.
All root spaces corresponding to non zero roots are of dimension one, and they are given by

(6) $[H, E_{i,j}] = (\lambda_i - \lambda_j)(H)\, E_{i,j}$ for all $H \in \underline{h}$.

Proof. The verification of (6) is elementary. The fact that there are no other roots than those written above follows from lemma 6 where $\underline{m}$ can be chosen as the space of all operators in $\underline{sl}(\mathcal{H}; C_o)$ whose matrix representations with respect to e have no diagonal terms, and where $\langle\langle\ |\ \rangle\rangle$ can be defined by $\langle\langle X \mid Y\rangle\rangle = \mathrm{trace}(XY^*)$. ■

In the case of proposition 4A, the c-involutive Lie algebra having a Cartan decomposition associated to $\underline{g}$ and $\underline{h}$ can be defined to be the derived algebra of the Lie algebra $\underline{h} \oplus (\oplus_{\alpha \in \mathcal{R}} \underline{g}_{\alpha})$; it is clearly equal to $\underline{sl}(\infty, \mathbf{C}) = \bigcup_{n \in N} \underline{sl}(n, \mathbf{C})$, where $\underline{sl}(n, \mathbf{C})$ is the subalgebra of $\underline{sl}(\mathcal{H}; C_o)$ consisting of those operators which map the span of $(e_o, e_1, \dots, e_{n-1})$ into itself and its orthogonal complement in $\mathcal{H}$ onto zero. About the relationship between these considerations and L*-algebras, see the project at the end of section 1.4.

<u>Proposition 4B(type one)</u>. Let $\underline{g} = \underline{o}(\mathcal{H}, J_{\mathbb{R}}; C_o)$, let $\underline{h}$ be a Cartan subalgebra of $\underline{g}$ of type one, and let $e = (e_n)_{n \in Z}$ be a $J_{\mathbb{R}}$-basis compatible with $\underline{h}$. For each $i \in N$, let $\varepsilon_i = \frac{1}{2}(E_{i,i} - E_{-i,-i})$ and let $\lambda_i : \underline{h} \to \mathbb{C}$ be defined by $\lambda_i(H) = \mathrm{trace}(H\varepsilon_i)$. Then

$$\begin{aligned}\mathcal{R} = \; & \{\lambda_i - \lambda_j \in \underline{h}^{dual} \mid i,j \in N^*, \; i \neq j\} \cup \\ & \{\pm(\lambda_i + \lambda_j) \in \underline{h}^{dual} \mid i,j \in N^*, \; i < j\} \cup \\ & \{\pm \lambda_i \in \underline{h}^{dual} \mid i \in N^*\} .\end{aligned}$$

All root spaces corresponding to non zero roots are of dimension one, and they are given by

$$(7) \quad [H, E_{i,j} - E_{-j,-i}] = (\lambda_i - \lambda_j)(H) \quad (E_{i,j} - E_{-j,-i})$$

$$(8) \quad [H, E_{j,-i} - E_{i,-j}] = (\lambda_i + \lambda_j)(H) \quad (E_{j,-i} - E_{i,-j})$$

$$(9) \quad [H, E_{-i,j} - E_{-j,i}] = -(\lambda_i + \lambda_j)(H) \quad (E_{-i,j} - E_{-j,i})$$

$$(10) \quad [H, E_{j,o} - E_{o,-j}] = \lambda_j(H) \quad (E_{j,o} - E_{o,-j})$$

$$(11) \quad [H, E_{o,j} - E_{-j,o}] = -\lambda_j(H) \quad (E_{o,j} - E_{-j,o})$$

for all $H \in \underline{h}$.

<u>Proof</u> : classical. The c-involutive Lie algebra having a Cartan decomposition associated to $\underline{g}$ and $\underline{h}$ is clearly $\underline{so}(\infty, \mathbb{C})$. ∎

<u>Proposition 4B(type two)</u>. Let $\underline{g} = \underline{o}(\mathcal{H}, J_{\mathbb{R}}; C_o)$, let $\underline{h}$ be a Cartan subalgebra of $\underline{g}$ of type two, and let $e = (e_n)_{n \in Z^*}$ be a $J_{\mathbb{R}}$-basis compatible with $\underline{h}$. For each $i \in N^*$, let ε_i and λ_i be defined formally as in proposition 4B(type one). Then

$$\begin{aligned}\mathcal{R} = \; & \{\lambda_i - \lambda_j \in \underline{h}^{dual} \mid i,j \in N^*, \; i \neq j\} \cup \\ & \{\pm(\lambda_i + \lambda_j) \; \underline{h}^{dual} \mid i,j \in N^*, \; i < j\} \cup\end{aligned}$$

All roots spaces corresponding to non zero roots are of dimension one, and they are given by

$$(12)\qquad [H, E_{i,j} - E_{-j,-i}] = (\lambda_i - \lambda_j)(H) \quad (E_{i,j} - E_{-j,-i})$$

$$(13)\qquad [H, E_{j,-i} - E_{i,-j}] = (\lambda_i + \lambda_j)(H) \quad (E_{j,-i} - E_{i,-j})$$

$$(14)\qquad [H, E_{-i,j} - E_{-j,i}] = -(\lambda_i + \lambda_j)(H) \quad (E_{-i,j} - E_{-j,i})$$

for all $H \in \underline{h}$.

Proof : as for proposition 4B(type one). ∎

Proposition 4C. Let $\underline{g} = \underline{sp}(\mathcal{H}, J_Q; C_o)$, let $\underline{h}$ be a Cartan subalgebra of $\underline{g}$ and let $e = (e_n)_{n \in Z^*}$ be a J_Q-basis compatible with $\underline{h}$. For each $i \in N^*$, let ε_i and λ_i be defined formally as in proposition 4B(type one). Then

$$= \{\lambda_i - \lambda_j \in \underline{h}^{dual} \mid i,j \in N^* , \ i \neq j\} \cup$$
$$\{\pm(\lambda_i + \lambda_j) \ \underline{h}^{dual} \mid i,j \in n^* , \ i < j\} \cup$$
$$\{ \pm 2 \lambda_i \ \underline{h}^{dual} \mid i \in N^*\} .$$

All root spaces corresponding to non zero roots are of dimension one, and they are given by

$$(15)\qquad [H, E_{i,j} - E_{-j,-i}] = (\lambda_i - \lambda_j)(H) \quad (E_{i,j} - E_{-j,-i})$$

$$(16)\qquad [H, E_{j,-i} + E_{i,-j}] = (\lambda_i + \lambda_j)(H) \quad (E_{j,-i} + E_{i,-j})$$

$$(17)\qquad [H, E_{-i,j} + E_{j,-i}] = -(\lambda_i + \lambda_j)(H) \quad (E_{-i,j} + E_{j,-i})$$

$$(18)\qquad [H , E_{i,-i}] = 2\lambda_i(H)\, E_{i,-i}$$

$$(19)\qquad [H , E_{-i,i}] = -2\lambda_i(H)\, E_{-i,i}$$

Proof : classical. The c-involutive Lie algebra having a Cartan decomposition associated to $\underline{g}$ and $\underline{h}$ is clearly $\underline{sp}(\infty , C)$. ∎

Let $\underline{g}$ be a classical complex Lie algebra of finite rank operators, let $\underline{h}$ be a Cartan subalgebra of $\underline{g}$ and let $\mathcal{R}$ be the set of non zero roots of $\underline{g}$ with respect to $\underline{h}$. Proposition 4 shows in particular that $\mathcal{R}$ behaves much as for finite dimensional classical algebras. Namely :

i) The set $\mathcal{R}$ is reduced : if $\alpha \in \mathcal{R}$, the only roots proportional to α are 0 and $\pm\alpha$.

ii) Let α,β be two roots in $\mathcal{R}$; then there exists two positive integers p and q such that, when $j \in Z$, $\beta + j\alpha$ is a root if and only if $-q \leq j \leq p$. The set $\{\gamma \in \mathcal{R} \cup \{0\} \mid \gamma = \beta + j\alpha$ for some $j \in Z\}$ is called the α-chain of roots defined by β, and $q + p$ is the length of the chain. The rational integer $q - p$ will be denoted by $n(\beta,\alpha)$. The quotient $n(\alpha,\beta)/n(\beta,\alpha)$, if it is defined, will be called the ratio of α and β, and will be denoted by $r(\alpha,\beta)$. Given $\alpha \in \mathcal{R}$, the maximum of the numbers $\sqrt{r(\alpha,\beta)}$ when β runs over all roots in $\mathcal{R}$ which make them defined will be called the length of the root α.

iii) Let α,β be two roots in $\mathcal{R}$. From the definition of $n(\beta,\alpha)$ given above, it is clear that $\beta - n(\beta,\alpha)\alpha$ is again a non zero root. The symmetry of root α is by definition the map

$$s_\alpha \begin{cases} \mathcal{R} \longrightarrow \mathcal{R} \\ \beta \longmapsto \beta - n(\beta,\alpha)\alpha \; ; \end{cases}$$

it is never the identity (because $n(\alpha,\alpha) = 2$) and $s_\alpha^{\;2} = id_{\mathcal{R}}$ (because $n(\beta - j\alpha,\ \alpha) = n(\beta,\alpha) - 2j$). The group of permutations of $\mathcal{R}$ generated by these symmetries is the Weyl group of g with respect to h .

Those complex c-involutive Lie algebras for which the above definitions make sense will be called admissible. Finite dimensional semi-simple complex Lie algebras are all admissible (see Serre [157] chap. V and VI, and/or Bourbaki [26] chap. VI §1, in particular proposition 9).

Definition 6. A semi-simple c-involutive complex Lie algebra g is said to be admissible if the following holds.

Let h be any Cartan subalgebra of g and let $\mathcal{R}$ be the set of non zero roots of $\underline{g}$ with respect to $\underline{h}$. Then :

i) $\mathcal{R}$ is reduced.

ii) Let $\alpha, \beta \in \mathcal{R}$; then there exists $p, q \in N$ such that, when $j \in Z$, $\beta + j\alpha \in \mathcal{R} \cup \{o\}$ if and only if $-q \leq j \leq p$.

iii) The length of each non zero root is finite.

iv) If the ratio $r(\alpha, \beta)$ as above is defined, then $\sqrt{r(\alpha, \beta)}$ is equal to the quotient of the length of α by that of β.

In an admissible Lie algebra, the ratio of any two roots can now be defined in the obvious way. In our standard examples, proposition 5 gives the values of the quantities defined above.

Proposition 5.

A. - Let $\underline{g} = \underline{sl}(\mathcal{H}; C_o)$, let $\underline{h}$ be a Cartan subalgebra of $\underline{g}$ and let $\mathcal{R}$ be the set of non zero roots of $\underline{g}$ with respect to $\underline{h}$. Let α and β be two roots in $\mathcal{R}$ which are not proportional. Then :

i) $n(\beta, \alpha) \in \{-1, 0, +1\}$;

ii) $n(\beta, \alpha) = 0$ if and only if the length of the α-chain of roots defined by β is zero ;

iii) $r(\alpha, \beta)$ is always equal to $+1$.

B(type one). - Let $\underline{g} = \underline{o}(\mathcal{H}, J_R; C_o)$, let $\underline{h}$ be a Cartan subalgebra of $\underline{g}$ of type one and let $\mathcal{R}$ be the set of non zero roots of $\underline{g}$ with respect to h. Let α and β be two roots in which are not proportional. Then :

i) $n(\beta, \alpha) \in \{-2, -1, 0, +1, +2\}$ and the five values do occur ;

ii) α and β can be chosen such that $n(\beta, \alpha) = 0$ and such that the length of the α-chain of roots defined by β is two ;

iii) $r(\alpha, \beta) \in \{\frac{1}{2}, 1, 2\}$ and the three values do occur.

B(type two). - Let $\underline{g} = \underline{o}(\mathcal{H}, J_{\mathbb{R}}; C_o)$, let $\underline{h}$ be a Cartan subalgebra of $\underline{g}$ of type two and let $\mathcal{R}$ be the set of non zero roots of $\underline{g}$ with respect to $\underline{h}$. Let α and β be two roots in $\mathcal{R}$ which are not proportional. Then :

i) $n(\beta, \alpha) \in \{-1, 0, +1\}$;

ii) $n(\beta, \alpha) = 0$ if and only if the length of the α-chain of roots defined by β is zero.

iii) $r(\alpha, \beta)$ is always equal to $+1$.

C. - Let $\underline{g} = \underline{sp}(\mathcal{H}, J_{\mathbb{Q}}; C_o)$, let $\underline{h}$ be a Cartan subalgebra of $\underline{g}$ and let $\mathcal{R}$ be the set of non zero roots of $\underline{g}$ with respect to $\underline{h}$. Let α and β be two roots in $\mathcal{R}$ which are not proportional. Then :

i) $n(\beta, \alpha) \in \{-2, -1, 0, 1, 2\}$ and the five values do occur ;

ii) $n(\beta, \alpha) = 0$ if and only if the length of the α-chain of roots defined by β is zero ;

iii) $r(\alpha, \beta) \in \{\frac{1}{2}, 1, 2\}$ and the three values do occur.

Proof : this follows from proposition 4 by a tedious but elementary inspection. ∎

Let $\underline{g_1}$ and $\underline{g_2}$ be two semi-simple c-involutive complex Lie algebras, let $\underline{h_1}$ be a Cartan subalgebra of $\underline{g_1}$ and let $\mathcal{R}_1$ be the set of non zero roots of $\underline{g_1}$ with respect to $\underline{h_1}$. Let $\varphi : \underline{g_1} \longrightarrow \underline{g_2}$ be an isomorphism for the structures of involutive complex Lie algebras, let $\underline{h_2} = \varphi(\underline{h_1})$ and let $\mathcal{R}_2 = \{\beta \in \underline{h_2}^{dual} \mid \beta = \alpha.\varphi^{-1}$ for some $\alpha \in R_1\}$. It is clear that $\underline{h_2}$ is a Cartan subalgebra of $\underline{g_2}$ and that $\mathcal{R}_2$ is the set of non zero roots of $\underline{g_2}$ with respect to $\underline{h_2}$ The map from $\mathcal{R}_1$ to $\mathcal{R}_2$ which sends α to $\alpha.\varphi^{-1}$ associates chains of roots to chains of roots ; moreover,
$n(\beta \circ \varphi^{-1}, \alpha \circ \varphi^{-1}) = n(\beta, \alpha)$ for all $\alpha, \beta \in \mathcal{R}_1$. In particular, the following proposition is a straightforward corollary of proposition 5.

Proposition 6.

i) The Lie algebras $\underline{sl}(\mathcal{H}; C_o)$, $\underline{o}(\mathcal{H}, J_{\mathbb{R}}; C_o)$ and $\underline{sp}(\mathcal{H}, J_{\mathbb{Q}}; C_o)$ are pairewise non-isomorphic as involutive complex Lie algebras.

ii) Let $\underline{h}$ be a Cartan subalgebra of $\underline{o}(\mathcal{H}, J_{\mathbb{R}}; C_o)$ and let φ be an automorphism of $\underline{o}(\mathcal{H}, J_{\mathbb{R}}; C_o)$ which commutes with the involution; then $\underline{h}$ and $\varphi(\underline{h})$ are of the same type.

Definition 7. Let $\underline{g}$ be a semi-simple c-involutive complex Lie algebra, let $\underline{h}$ be a Cartan subalgebra of $\underline{g}$ and let $\mathcal{R}$ be the set of non zero roots of $\underline{g}$ with respect to $\underline{h}$. A simple basis of roots in $\mathcal{R}$ is a subset S of $\mathcal{R}$ such that

i) the roots in S are linearly independent in $\underline{h}^{dual}$,

ii) any root in $\mathcal{R}$ can be written as a linear combination of elements in S, with integer coefficients which are all of the same sign.

Let $\underline{g}$, $\underline{h}$ and $\mathcal{R}$ be as in definition 7, let $\underline{k}$ be the subpsace of $\underline{h}^{dual}$ span by $\mathcal{R}$ and let S be a simple basis of roots in $\mathcal{R}$. Positive roots, simple roots and the lexicographic order of $\underline{k}$ given by a total order on S are defined as in the finite dimensional case. Chosen an order, the set of positive [resp. negative] roots in $\mathcal{R}$ will be denoted by $\mathcal{R}^+$ [resp. $\mathcal{R}^-$]. It is not clear (to me) which admissible semi-simple c-involutive complex Lie algebra admit simple basis of roots. Following Balachandran [11], such an algebra is said to be regular if it does admit such a basis of roots for any Cartan subalgebra.

Proposition 7A. With the same notations as in proposition 5A, let $\alpha_i = \lambda_i - \lambda_{i+1}$ and let s_i be the symmetry of root α_i, for each $i \in N$. Then

i) $S = (\alpha_i)_{i \in N}$ is a simple basis of roots in $\mathcal{R}$.

ii) The Weyl group of $\underline{g}$ with respect to $\underline{h}$, say W_A, is generated by $(s_i)_{i \in N}$. In other words, W_A is isomorphic to the group of finite permutations of the set N.

Proof : again by inspection, from proposition 4A; the consideration of subsystems of roots and of ad hoc subalgebras (as in Schue [153] section 3) can shorten the inspection needed for ii). ■

The reader will easily guess by now what propositions 7B(type one), 7B(type two) and 7C are. The only point to stress is that a Weyl group is attached to a Lie algebra given together with a type of Cartan subalgebra of it.

Project : The conditions imposed on the involution and the system of roots in definition 2 and 6 seem very restrictive. As a starting point for a general study, the following can be proposed :

Conjecture : Let $\underline{g}$ be a regular semi-simple c-involutive complex Lie algebra; then there exists a semi-simple complex L*-algebra $\bar{\underline{g}}$ which contains $\underline{g}$ as a self-adjoint dense subalgebra.

If true, this conjecture would be a sort of analogue of the results due to Kaplansky, about the classification of the semi-simple associative algebras known as semi-simple dual Q-rings [97].

I.5. - Automorphisms

As in sections I.3 and I.4, $\mathcal{H}$ is supposed to be an <u>infinite dimensional separable complex Hilbert space</u> in this section. Up to the notations, results would be the same for arbitrary dimension.

<u>Proposition 8</u>. Let $\underline{g}$ be one of the c-involutive complex Lie algebras $\underline{sl}(\mathcal{H} ; C_o)$, $\underline{o}(\mathcal{H} ,J_{\mathbb{R}}; C_o)$ and $\underline{sp}(\mathcal{H} ,J_{\mathbb{Q}}; C_o)$ defined in section I.1. Then there exists a hermitian form $B : \underline{g} \times \underline{g} \longrightarrow \mathbb{C}$ such that

(20) $B([X,Y] \mid Z) = B(Y \mid [X^*,Z])$ for all $X,Y,Z \in \underline{g}$. Any hermitian form with this property is a scalar multiple of the scalar product $\langle\langle \mid \rangle\rangle \begin{cases} \underline{g} \times \underline{g} & \longrightarrow \mathbb{C} \\ (X,Y) & \longmapsto \text{trace}(XY^*) \end{cases}$.

Let now $\underline{h}$ be a Cartan subalgebra of $\underline{g}$ and let $\mathcal{R}$ be the set of non zero roots of $\underline{g}$ with respect to $\underline{h}$. For any $\alpha \in \mathcal{R}$, there exists a unique vector $H_\alpha \in \underline{h}$ such that $\alpha(H) = \langle\langle H \mid H_\alpha \rangle\rangle$ for all $H \in \underline{h}$. Moreover,

$$(21) \qquad r(\alpha, \beta) = \frac{\langle\langle H_\alpha \mid H_\alpha \rangle\rangle}{\langle\langle H_\beta \mid H_\beta \rangle\rangle} \qquad \text{for all } \alpha, \beta \in \mathcal{R} .$$

<u>Proof</u> : The sesquilinear form $\langle\langle \mid \rangle\rangle$ defined in the proposition is clearly hermitian definite positive, and it satisfies (20); hence the existence. Let now B be a hermitian form on $\underline{g}$ satisfying (20). For any finite dimensional subspace E of $\mathcal{H}$, let B_E be the restriction of B to the subalgebra of $\underline{g}$ consisting of those operators which map E into itself and its orthogonal onto zero (E is submitted to the same restrictions as in lemma 3, page I.6). According to well-known results (see for example Koszul [100] théorèmes 11.1 and 11.2), there exists a family (c_E) of complex numbers such that $B_E(X \mid Y) = c_E \text{trace}(XY^*)$ for all $X,Y \in \underline{g} \cap \underline{end}(E)$. As the B_E's are

restrictions of each other, the c_E's must be equal to a same constant c. Hence the unicity.

The last part of proposition 8 follows from a contemplation of the facts collected in proposition 4 and 5. ∎

N.B. : We will sometimes write "the root H_α" instead of "the vector H_α in $\underline{h}$ which represents the root α of $\underline{g}$ with respect to $\underline{h}$ in the sense of proposition 8".

Proposition 9. Let $\underline{g}$ and $\langle\langle\,|\,\rangle\rangle$ be as in proposition 8 and let φ be a *-automorphism of $\underline{g}$. Then φ is unitary :

(22) $\langle\langle \varphi(X) \mid \varphi(Y) \rangle\rangle = \langle\langle X \mid Y \rangle\rangle$ for all $X,Y \in \underline{g}$.

Proof. It follows trivially from the unicity part of proposition 8 that there exists a non zero constant $c \in \mathbf{C}^*$ such that

(23) $\langle\langle \varphi(X) \mid \varphi(Y) \rangle\rangle = c\langle\langle X \mid Y\rangle\rangle$ for all $X,Y \in \underline{g}$. Let now $\underline{h}$ be a Cartan subalgebra of $\underline{g}$, let α be a non zero root of $\underline{g}$ with respect to $\underline{h}$, let H_α be as in proposition 8, and let X_α be a non zero root vector of α, so that

(24) $[H,X_\alpha] = \langle\langle H \mid H_\alpha\rangle\rangle X_\alpha$ for all $H \in \underline{h}$. Let $\underline{k} = \varphi(\underline{h})$; then $\underline{k}$ is a Cartan subalgebra of $\underline{g}$ (of the same type as $\underline{h}$ if $\underline{g} = \underline{o}(\mathcal{H}, J_R;\, C_o)$), $\alpha\circ\varphi^{-1}$ is a non zero root of $\underline{g}$ with respect to $\underline{k}$, and $\varphi(X_\alpha)$ is a non zero root vector of $\alpha\circ\varphi^{-1}$; let K_α be the vector of $\underline{k}$ such that $\langle\langle K \mid K_\alpha\rangle\rangle = \alpha\circ\varphi^{-1}(K)$ for all $K \in \underline{k}$. By application of φ to the equality (24), and then by use of (23):

$$[K, \varphi(X_\alpha)] = \frac{1}{c} \langle\langle K \mid \varphi(H_\alpha)\rangle\rangle \varphi(X_\alpha) \quad \text{for all } K \in \underline{k}.$$

But on the other hand :

$$[K, \varphi(X_\alpha)] = \langle\langle K \mid K_\alpha\rangle\rangle \varphi(X_\alpha) \quad \text{for all } K \in \underline{k}.$$

It follows that $K_\alpha = \frac{1}{c}\varphi(H_\alpha)$. Now, clearly, $\langle\langle H_\alpha \mid H_\alpha \rangle\rangle =$ $= \langle\langle K_\alpha \mid K_\alpha\rangle\rangle$: hence, using (23) :

$$\langle\langle H_\alpha \mid H_\alpha \rangle\rangle = \frac{1}{c\bar{c}} \langle\langle \varphi(H_\alpha) \mid \varphi(H_\alpha) \rangle\rangle = \frac{1}{\bar{c}} \langle\langle H_\alpha \mid H_\alpha \rangle\rangle \quad ,$$

so that $c = 1$. ∎

I.5A. - *-automorphisms of $\underline{g} = \underline{sl}(\mathcal{H};\, C_o)$

Let φ be a *-automorphism of $\underline{g}$, let $\underline{h}$ be a Cartan subalgebra of $\underline{g}$, let $\underline{k}$ be the Cartan subalgebra $\varphi(\underline{h})$ of $\underline{g}$, and let $e = (e_n)_{n \in N}$ [resp. $f = (f_n)_{n \in N}$] be an orthonormal basis of $\mathcal{H}$ compatible with $\underline{h}$ [resp. with $\underline{k}$] ; let $(E_{i,j})_{i,j \in N}$ [resp. $(F_{i,j})_{i,j \in N}$] be the corresponding operators defined as before proposition 4A; for each $i \in N$, let E_i denote $E_{i,i}$ and let F_i denote $F_{i,i}$.

Lemma 7A. Let $J_{\mathbb{R}}$ be the conjugation in $\mathcal{H}$ such that $J_{\mathbb{R}} e_n = e_n$ for all $n \in N$. Then there exists a unitary operator U such that φ coincides on $\underline{h}$

either with $\quad \varphi_{UJ} : H \longmapsto -UJ_{\mathbb{R}}H^*J_{\mathbb{R}}U^*$

or with $\quad \varphi_U : H \longmapsto UHU^*$.

Proof.

φ induces a bijective map from the roots of $\underline{g}$ with respect to $\underline{h}$ to the roots of $\underline{g}$ with respect to $\underline{k}$, hence from the vectors in $\underline{h}$ representing the roots of $\underline{g}$ with respect to $\underline{h}$ (the H_α's of proposition 8) to the vectors in $\underline{k}$ representing the roots of $\underline{g}$ with respect to $\underline{k}$. Hence, there exist two maps $a,b : N \to N$ such that $a(n) \neq b(n)$ for all $n \in N$ and such that (see proposition 4A)

$$\varphi(E_n - E_{n+1}) = F_{a(n)} - F_{b(n)} \quad \text{for all } n \in N .$$

As $E_o - E_2$ is a root with respect to $\underline{h}$,
$\varphi(E_o - E_2) = F_{a(o)} - F_{b(o)} + F_{a(1)} - F_{b(1)}$ must be a root with

respect to $\underline{k}$, so that either $F_{a(o)} = F_{b(1)}$, or $F_{b(o)} = F_{a(1)}$.

Suppose first that $a(o) = b(1)$. As $E_1 - E_3$ is a root with respect to $\underline{h}$, $\varphi(E_1 - E_3) = F_{a(1)} - F_{b(1)} + F_{a(2)} - F_{b(2)}$ is a root with respect to $\underline{k}$. If one had $F_{b(1)} = F_{a(2)}$, this would imply the clearly absurd relation $\varphi(E_o-E_3)=F_{b(1)}+F_{a(1)}-F_{b(o)}-F_{b(2)}$; hence $F_{a(1)} = F_{b(2)}$. By iteration of the same argument : $\varphi(E_n - E_{n+1}) = F_{b(n+1)} - F_{b(n)}$ for all $n \in N$ and the map b must be a permutation of N . By reordering the basis f , we can suppose that $\varphi(E_n - E_{n+1}) = F_{n+1} - F_n$ for all $n \in N$. Let now U be the unitary operator on $\mathcal{H}$ such that $Ue_n = f_n$ for all $n \in N$ and let φ_{UJ} be the *-automorphism $\begin{cases} \underline{g} \longrightarrow \underline{g} \\ X \longmapsto -UJ_{\mathbb{R}}X^*J_{\mathbb{R}}U^* \end{cases}$. As $\varphi_{UJ}(E_n) = -F_n$ for all $n \in N$, φ and φ_{UJ} coincide on the set $(E_n-E_{n+1})_{n \in N}$ which generate $\underline{h}$, hence φ and φ_{UJ} coincide on $\underline{h}$.

Suppose now that $b(o) = a(1)$. The same argument implies that modulo a reordering of the basis f , $\varphi(E_n - E_{n+1}) = F_n - F_{n+1}$ for all $n \in N$. Hence φ coincides on $\underline{h}$ with the restriction of the *-automorphism

$\begin{cases} \underline{g} \longrightarrow \underline{g} \\ X \longmapsto UXU^* \end{cases}$, where U is the unitary operator on $\mathcal{H}$ such that $Ue_n = f_n$ for all $n \in N$. ∎

<u>Lemma 8A1</u>. Suppose that φ coincides on $\underline{h}$ with the *-automorphism φ_{UJ} . Then there exists a unitary operator V on $\mathcal{H}$ such that $\varphi = \varphi_{VJ}$.

<u>Proof</u>.

The root vector corresponding to the root $E_p - E_q$ of $\underline{g}$ with respect to $\underline{h}$ is $E_{p,q}$. Hence $\varphi(E_{p,q})$ must be a root vector corresponding to the root $F_q - F_p$ of $\underline{g}$ with respect to $\underline{k}$. As φ

is unitary (proposition I.9), there exists a complex number of modulus one, say $\exp(is_{p,q})$ with $s_{p,q} \in \mathbb{R}$, such that $\varphi(E_{p,q}) = \exp(is_{p,q})F_{q,p}$. This can be done for all $p,q \in N$. As $E_{p,q}^* = E_{q,p}$ one must have $\exp(-is_{p,q}) = \exp(is_{q,p})$ for all $p,q \in N$. As $[E_{p,n}, E_{n,q}] = E_{p,q}$ one must have $\exp(is_{p,n} + is_{n,q}) = -\exp(is_{p,q})$ for all triples p,n,q of distinct positive integers.

Let S be the operator on $\mathcal{H}$ whose representation with respect to the basis f is the matrix

$$\begin{pmatrix} 1 & & & & & 0 \\ & \exp(is_{o,1}) & & & & \\ & & \exp(is_{o,2}) & & & \\ & & & \exp(is_{o,3}) & & \\ & 0 & & & \ddots & \\ & & & & & \ddots \end{pmatrix}$$

and let φ_{SUJ} be the $*$-automorphism $\begin{cases} g \longrightarrow g \\ X \mapsto -SUJ_{\mathbb{R}}X^*J_{\mathbb{R}}U^*S^* \end{cases}$.

Then $\varphi_{SUJ}(E_n) = +S\ \varphi_{UJ}(E_n)S^* = -F_n = \varphi(E_n)$ for all $n \in N$ and

$\varphi_{SUJ}(E_{p,q}) = +S\ \varphi_{UJ}(E_{p,q})S^* = -\ SF_{q,p}S^* =$

$= -\exp(is_{o,q} - is_{o,p})F_{q,p} = +\exp(is_{p,q})F_{q,p} = \varphi(E_{p,q})$ for all $p,q \in N$ with $p \neq q$. As the E_n's and the $E_{p,q}$'s generate a subspace dense in g (with respect to $\langle\langle\ |\ \rangle\rangle$), φ_{SUJ} is identical to φ, which ends the proof. ■

<u>Lemma 8A2</u>. Suppose that φ coincides on $\underline{h}$ with the $*$-automorphism φ_U. Then there exists a unitary operator V on $\mathcal{H}$ such that $\varphi = \varphi_V$.

<u>Proof</u> : as for 8A1. ■

<u>Proposition 10A</u>. Let $g = \underline{sl}(\mathcal{H}; C_o)$ and let $J_{\mathbb{R}}$ be a fixed conjugation on $\mathcal{H}$. Let φ be a $*$-automorphism of g. Then there

exists a unitary operator $V \in U(\mathcal{H})$ such that

$$\text{either} \quad \varphi = \varphi_{VJ} \qquad \begin{cases} \mathfrak{g} \longrightarrow \mathfrak{g} \\ X \longmapsto -VJ_{\mathbb{R}}X^*J_{\mathbb{R}}V^* \end{cases}$$

$$\text{or} \quad \varphi = \varphi_V \qquad \begin{cases} \mathfrak{g} \longrightarrow \mathfrak{g} \\ X \longmapsto VXV^* \end{cases} .$$

The two cases exclude each other.

The operator V is uniquely determined by φ, up to multiplication by a complex number of modulus one.

Otherwise said, the sequence

$$\{1\} \longrightarrow U(1) \xrightarrow{\ j\ } \tilde{U}(\mathcal{H}) \xrightarrow{\ \pi\ } \overset{*}{\mathrm{Aut}}(\mathfrak{g}) \longrightarrow \{1\}$$

is exact, where $\tilde{U}(\mathcal{H})$ is the group of all unitary and antiunitary bounded operators on $\mathcal{H}$, and where $\overset{*}{\mathrm{Aut}}(\mathfrak{g})$ is the group of $*$-automorphisms of $\mathfrak{g}$; the map j is given by $j(e^{i\psi}) = e^{i\psi}\,\mathrm{id}_{\mathcal{H}}$, and the map π is given by $\pi(V) = \varphi_V$ [resp. $\pi(VJ_{\mathbb{R}}) = \varphi_{VJ}$] when V is unitary [resp. when $VJ_{\mathbb{R}}$ is antiunitary].

Proof.

The existence of the operator V has been proved in lemmas 7A and 8A.

Suppose now that $\varphi_{V_1} = \varphi_{V_2J}$ for some unitary operators V_1, V_2 on $\mathcal{H}$ and let $V = V_2^*V_1$; then $VXJ_R = -X^*J_{\mathbb{R}}V$ for all $X \in \mathfrak{g}$. Let y be a vector of norm 1 in $\mathcal{H}$; for each vector z in $\mathcal{H}$ which is of norm 1 and orthogonal to y, $y \otimes \bar{z} \in \mathfrak{g}$, so that $V(y \otimes \bar{z})J_{\mathbb{R}} = -(z \otimes \bar{y})J_{\mathbb{R}}V$ and $Vy = (V(y \otimes \bar{z})J_{\mathbb{R}})(J_{\mathbb{R}}z) = (-(z \otimes \bar{y})J_{\mathbb{R}}V)(J_{\mathbb{R}}z) = -\langle J_{\mathbb{R}}VJ_{\mathbb{R}}z \mid y\rangle z$. It follows that $Vy = 0$, which is absurd because V is unitary.

The third part of proposition 10A is a straightforward consequence of Schur's lemma. ∎

Remarks.

i) Proposition 10A is still true if $\underline{sl}(\mathcal{H}, C_o)$ is replaced by $\underline{gl}(\mathcal{H}, C_o)$. In other words, if φ is a *-automorphism of $\underline{gl}(\mathcal{H}; C_o)$, then φ is either a *-automorphism of the associative simple algebra $C_o(\mathcal{H})$, or the negative of a *-antiautomorphism of $C_o(\mathcal{H})$. This is the generalisation of a well-known result about rings of matrices, and should be compared with general results obtained more recently by Martindale [118].

ii) Let D be a bounded skew-adjoint operator on $\mathcal{H}$ which has a purely continuous spectrum and such that $\| D \| < \frac{1}{2}\log 2$. Let φ be the *-automorphism of $\underline{sl}(\mathcal{H}; C_o)$ defined by $\varphi(X) = \exp(D)\, X \exp(-D)$. Then φ has no non zero fixed point [13]. Such fixed-point-free automorphisms do not occur in finite dimensions (see Borel-Mostow [23]).

I.5B. - *-automorphisms of $\underline{g} = \underline{o}(\mathcal{H}, J_{\mathbb{R}}; C_o)$

Let φ be a *-automorphism of $\underline{g}$, let $\underline{h}$ be a Cartan subalgebra of $\underline{g}$ of type two, let $\underline{k}$ be the Cartan subalgebra $\varphi(\underline{h})$ of $\underline{g}$, which is of type two (proposition I.6), and let $e = (e_n)_{n \in Z^*}$ [resp. $f = (f_n)_{n \in Z^*}$] be an orthonormal basis of $\mathcal{H}$ compatible with $\underline{h}$ [resp. with $\underline{k}$]; let $(E_{i,j})_{i,j \in Z^*}$ [resp. $(F_{i,j})_{i,j \in Z^*}$] be the corresponding elementary operators; for each $i \in N^*$, let ε_i denote $\frac{1}{2}(E_{i,i} - E_{-i,-i})$ and let $\mathcal{F}_i$ denote $\frac{1}{2}(F_{i,i} - F_{-i,-i})$.

Lemma 7B. There exists a unitary operator U on $\mathcal{H}$ which commutes with $J_{\mathbb{R}}$ and such that φ coincides on $\underline{h}$ with $\varphi_U : H \longmapsto UHU^*$.

Proof.

For the same reasons as in lemma 7A, there exist two maps $a,b : N^* \longrightarrow N^*$ such that $a(n) \neq b(n)$ for all $n \in N^*$ and two sequences $c = (c_n)_{n \in N^*}$, $d = (d_n)_{n \in N^*}$ consisting of zeros and ones only, such that

(25) $\varphi(\varepsilon_n - \varepsilon_{n+1}) = (-1)^{c_n} \mathcal{F}_{a(n)} + (-1)^{d_n} \mathcal{F}_{b(n)}$ for all $n \in N^*$.

As $\varepsilon_1 - \varepsilon_3$ is a root with respect to $\underline{h}$, $\varphi(\varepsilon_1 - \varepsilon_3) =$

$$= (-1)^{c_1} \mathcal{F}_{a(1)} + (-1)^{d_1} \mathcal{F}_{b(1)} + (-1)^{c_2} \mathcal{F}_{a(2)} + (-1)^{d_2} \mathcal{F}_{b(2)}$$

must be a root with respect to $\underline{k}$. It follows that, possibly modulo a re-definition of a, b, c and d, equation (25) can be written for $n = 1$ and for $n = 2$ as

$$\varphi(\varepsilon_1 - \varepsilon_2) = (-1)^{c_1} \mathcal{F}_{a(1)} - (-1)^{c_2} \mathcal{F}_{a(2)}$$

$$\varphi(\varepsilon_2 - \varepsilon_3) = (-1)^{c_2} \mathcal{F}_{a(2)} + (-1)^{d_2} \mathcal{F}_{b(2)} .$$

By iteration of the same procedure, one obtains a map $a : N^* \longrightarrow N^*$ which must be a permutation of N^*, and a sequence $c = (c_n)_{n \in N^*}$ consisting of zeros and ones only, such that

(26) $\varphi(\varepsilon_n - \varepsilon_{n+1}) = (-1)^{c_n} \mathcal{F}_{a(n)} - (-1)^{c_{n+1}} \mathcal{F}_{a(n+1)}$ for all $n \in N^*$.

Let now $f' = (f'_n)_{n \in Z^*}$ be the orthonormal basis of defined by

$$\begin{cases} f'_n = f_{n'} \\ f'_{-n} = f_{-n'} \end{cases} \quad \text{where } n' = (-1)^{c_n} a(n) \quad \text{for all } n \in N^*$$

and let $(\mathcal{F}'_n)_{n \in N^*}$ be the corresponding sequence of operators :

$$\mathcal{F}'_n = (-1)^{c_n} \mathcal{F}_{a(n)} \qquad \text{for all } n \in N^* .$$

The basis f' of $\mathcal{H}$ is clearly compatible with $\underline{k}$, and equations (26) can now be written as $\varphi(\varepsilon_n - \varepsilon_{n+1}) = \mathcal{F}'_n - \mathcal{F}'_{n+1}$ for all $n \in N^*$.

It follows that there is no loss of generality in supposing to start with that (26) can be written as

(27) $\varphi(\varepsilon_n - \varepsilon_{n+1}) = \mathcal{F}_n - \mathcal{F}_{n+1}$ for all $n \in N^*$.

The same argument as that used to end the proof of lemma 7A can now be applied. ∎

Lemma 8B. There exists a unitary operator V on $\mathcal{H}$ which commutes with $J_{\mathbb{R}}$ and such that $\varphi = \varphi_V$.

Proof.

For reasons analogous to those in the proof of lemma 8A, there exist three sequences of real numbers $(s_{p,q})_{p,q \in N^*}$, $(t_{p,q})_{p,q \in N^*}$ and $(v_{p,q})_{p,q \in N^*}$, with $t_{p,q} = t_{q,p}$ and $v_{p,q} = v_{q,p}$ for all $p,q \in N^*$, $p > q$, such that :

$$\varphi(E_{p,q} - E_{-q,-p}) = \exp(is_{p,q})\,(F_{p,q} - F_{-q,-p})$$
$$\varphi(E_{q,-p} - E_{p,-q}) = \exp(it_{p,q})\,(F_{q,-p} - F_{p,-q})$$
$$\varphi(E_{-p,q} - E_{-q,p}) = \exp(iv_{p,q})\,(F_{-p,q} - F_{-q,p})$$

for all $p,q \in N^*$, $p \neq q$.

Again as in lemma 8A : $\exp(-is_{p,q}) = \exp(is_{q,p})$ and $\exp(-it_{p,q}) = \exp(iv_{p,q})$ for all $p,q \in N^*$, $p \neq q$; $\exp(is_{p,n} + is_{n,q}) = \exp(is_{p,q})$ and $\exp(it_{p,m} + iv_{m,q}) = \exp(is_{p,q})$ for all triple of distinct natural integers p,m,q.

Let S be the operator on $\mathcal{R}$ whose representation with respect to the basis f is the matrix

$$\left(\begin{array}{ccc|ccc} \ddots & & & & & \\ & e^{is_{1,3}} & & & 0 & \\ & & e^{is_{1,2}} & & & \\ & & 1 & & & \\ \hline & & & 1 & & \\ & & & & e^{-is_{1,2}} & \\ & 0 & & & e^{-is_{1,3}} & \\ & & & & & \ddots \end{array}\right)$$

Then S commutes with $J_{\mathbb{R}}$, and :

$$\varphi_{SU}(\varepsilon_n) = S\,\mathcal{F}_n S^* = \mathcal{F}_n \quad \text{for all} \quad n \in N^*$$

$$\varphi_{SU}(E_{p,q} - E_{-q,-p}) = \exp(-is_{1,p} + is_{1,q})\,(F_{p,q} - F_{-q,-p}) =$$

$$= \varphi(E_{p,q} - E_{-q,-p})$$

$$\varphi_{SU}(E_{q,-p} - E_{p,-q}) = \exp(-is_{1,p} - is_{1,q})\,(F_{q,-p} - F_{p,-q})$$

$$\varphi_{SU}(E_{-p,q} - E_{-q,p}) = \exp(is_{1,p} + is_{1,q})\,(F_{-p,q} - F_{-q,p})$$

for all $p,q \in N^*$ with $p \neq q$.

If $w_{p,q} = s_{1,p} + s_{1,q} + t_{p,q}$ for all $p,q \in N^*$ with $p \neq q$, it is easy to check that $\exp(iw_{p,q})$ takes the same value, say $\exp(iw)$ for all $p,q \in N^*$, $p \neq q$. It follows that the two last expressions can be written as

$$\varphi_{SU}(E_{q,-p} - E_{p,-q}) = \exp(-iw)\quad(F_{q,-p} - F_{p,-q})$$

$$\varphi_{SU}(E_{-p,q} - E_{-q,p}) = \exp(iw)\quad(F_{-p,q} - F_{-q,p})$$

for all $p,q \in N^*$ with $p \neq q$.

Let finally T be the operator on $\mathcal{H}$ which is multiplication by $\exp(iw)$ in the span of $(f_n)_{n \in N^*}$ and multiplication by $\exp(-iw)$ in the span of $(f_{-n})_{n \in N^*}$. Then TSU commutes with $J_{\mathbb{R}}$ and $\varphi = \varphi_{TSU}$. ∎

<u>Proposition 10B</u>. Let $\underline{g} = \underline{o}(\mathcal{H}, J_{\mathbb{R}};\, C_o)$. Let φ be a *-automorphism of $\underline{g}$. Then there exists an operator $V \in O(\mathcal{H}_{\mathbb{R}}) = \{X \in U(\mathcal{H}) \mid XJ_{\mathbb{R}} = J_{\mathbb{R}}X\}$ such that

$$\varphi = \varphi_V \quad : \quad \begin{cases} \underline{g} \longrightarrow \underline{g} \\ X \longmapsto VXV^* \end{cases} .$$

The operator V is uniquely determined by φ , up to a sign. Otherwise said, the sequence

$$\{1\} \longrightarrow Z_2 \xrightarrow{\ j\ } O(\mathcal{H}_{\mathbb{R}}) \xrightarrow{\ \pi\ } \overset{*}{\mathrm{Aut}}(\underline{g}) \longrightarrow \{1\}$$

is exact, where the notations are similar to those defined in proposition 10A.

<u>Proof</u> : via lemmas 7B and 8B, and Schur's lemma. ∎

<u>Remarks</u>.

i) Any *-automorphism of $\underline{o}(\mathcal{H}, J_{\mathbb{R}};\, C_o)$ extends to a *-automorphism of $\underline{sl}(\mathcal{H};\, C_o)$, and this in two different ways. Indeed, both φ_V and φ_{VJ} restrict to

$$\begin{cases} \underline{o}(\mathcal{H}, J_{\mathbb{R}};\, C_o) \longrightarrow \underline{o}(\mathcal{H}, J_{\mathbb{R}};\, C_o) \\ X \longmapsto VXV^* = -\, VJ_{\mathbb{R}}X^*J_{\mathbb{R}}V^* \end{cases} .$$

ii) In finite dimension, it is well known that all automorphisms of $B_n = \underline{so}(2m+1, \mathbb{C})$ are inner. Inner automorphisms form a subgroup of index 2 in the group of automorphisms of $D_n = \underline{so}(2n, \mathbb{C})$ (if $n > 4$).

I.5C. - *-automorphisms of $\underline{g} = \underline{sp}(\mathcal{H}, J_Q; C_o)$

Let φ be a *-automorphism of $\underline{g}$, let $\underline{h}$ be a Cartan subalgebra of $\underline{g}$, let $\underline{k}$ be the Cartan subalgebra $\varphi(\underline{h})$ of $\underline{g}$, and let $e = (e_n)_{n \in Z^*}$ [resp. $f = (f_n)_{n \in Z^*}$] be an orthonormal basis of $\mathcal{H}$ compatible with $\underline{h}$ [resp. $\underline{k}$]; the notations $E_{i,j}$ $F_{i,j}$ ε_i $\mathcal{F}_i$ are used as before (page I.31).

Lemma 7C. There exists a unitary operator U on $\mathcal{H}$ which commutes with J_Q and such that φ coincides on $\underline{h}$ with $\varphi_U : H \longmapsto UHU^*$.

Proof.

φ induces a bijective map from the roots of $\underline{g}$ with respect to $\underline{h}$ which are of length $\sqrt{2}$ to the roots of $\underline{g}$ with respect to $\underline{k}$ which are of length $\sqrt{2}$. Hence there exists a map $a : N^* \longrightarrow N^*$ and a sequence $c = (c_n)_{n \in N^*}$ consisting of zeros and ones only, such that

(28) $\varphi(\varepsilon_n) = (-1)^{c_n} \mathcal{F}_{a(n)}$ for all $n \in N^*$.

Let now $f' = (f'_n)_{n \in Z^*}$ be the orthonormal basis of $\mathcal{H}$ defined by

$$\begin{cases} f'_n = (-1)^{c_n} f_{n'} \\ f'_{-n} = f_{-n'} \end{cases} \quad \text{where } n' = (-1)^{c_n} a(n) \quad \text{for all } n \in N^* .$$

and let $(\mathcal{F}'_n)_{n \in N^*}$ be the corresponding sequence of operators :

$$\mathcal{F}'_n = (-1)^{c_n} \mathcal{F}_{a(n)} \qquad \text{for all } n \in N^* .$$

The basis f' of $\mathcal{H}$ is clearly compatible with $\underline{k}$, and equations (28) can now be written as

$$\varphi(\varepsilon_n) = \mathcal{F}'_n \quad \text{for all } n \in N^* .$$

It follows that there is no loss of generality in supposing to start with that (28) can be written as

(29) $\varphi(\varepsilon_n) = \mathfrak{F}_n$ for all $n \in N^*$.

As for lemmas 7A and 7B, this ends the proof. ∎

<u>Lemma 8C</u>. There exists a unitary operator V on $\mathfrak{H}$ which commutes with J_Q and such that $\varphi = \varphi_V$.

<u>Proof</u>.

For reasons analogous to those in the proofs of lemmas 8A and 8B, there exists three sequences of real numbers

$(r_n)_{n \in N^*}$, $(s_{p,q})_{p,q \in N^*}$ and $(t_{p,q})_{p,q \in N^*}$, with $t_{p,q} = t_{q,p}$ for all $p,q \in N^*$, $p > q$, such that

$$\left.\begin{aligned} \varphi(E_{n,-n}) &= \exp(i2r_n)F_{n,-n} \\ \varphi(E_{-n,n}) &= \exp(-i2r_n)F_{-n,n} \end{aligned}\right\} \quad \text{for all } n \in N^*$$

$$\begin{aligned} \varphi(E_{p,q} - E_{-q,-p}) &= \exp(is_{p,q})\,(F_{p,q} - F_{-q,-p}) \\ \varphi(E_{q,-p} + E_{p,-q}) &= \exp(it_{p,q})\,(F_{q,-p} + F_{p,-q}) \\ \varphi(E_{-p,q} + E_{-q,p}) &= \exp(-it_{p,q})\,(F_{-p,q} + F_{-q,p}) \end{aligned}$$

$$\text{for all } p,q \in N^* , \ p \neq q .$$

Again as in lemma 8A : $\exp(-is_{p,q}) = \exp(is_{q,p})$ for all $p,q \in N^*$, $p \neq q$; $\exp(-i2r_p + is_{p,q}) = \exp(-it_{p,q})$ and $\exp(-i2r_q - is_{p,q}) = \exp(-it_{p,q})$ for all $p,q \in N^*$, $p \neq q$; the last two expressions can be written as

$$\left\{\begin{aligned} \exp(is_{p,q}) &= \exp(ir_p - ir_q) \\ \exp(it_{p,q}) &= \exp(ir_p + ir_q) \end{aligned}\right. \quad \text{for all } p,q \in N^* , \ p \neq q .$$

Let S be the operator on $\mathfrak{H}$ whose representation with respect to the basis f is the matrix

$$\left[\begin{array}{ccc|ccc} \ddots & & & & & \\ & e^{-ir_3} & & & 0 & \\ & & e^{-ir_2} & & & \\ & & \quad e^{-ir_1} & & & \\ \hline & & & e^{ir_1} & & \\ & & & & e^{ir_2} & \\ & 0 & & & & e^{ir_3} \\ & & & & & \ddots \end{array}\right]$$

Then S commutes with J_Q and $\varphi = \varphi_{SU}$. ∎

Proposition 10C. Let $\underline{g} = \underline{sp}(\mathcal{H}, J_Q; C_o)$. Let φ be a *-automorphism of $\underline{g}$. Then there exists an operator $V \in Sp(\mathcal{H}_Q) = \{X \in U(\mathcal{H}) \mid XJ_Q = J_Q X\}$ such that

$$\varphi = \varphi_V \;:\; \begin{cases} \underline{g} \longrightarrow \underline{g} \\ X \longmapsto VXV^* \end{cases} .$$

The operator V is uniquely determined by φ , up to a sign. Otherwise said, the sequence

$$\{1\} \to Z_2 \xrightarrow{\;j\;} Sp(\mathcal{H}_Q) \xrightarrow{\;\pi\;} \overset{*}{Aut}(\underline{g}) \to \{1\}$$

is exact, where the notations are similar to those defined in proposition 10A.

Proof : via lemmas 7C and 8C, and Schur's lemma. ∎

Remarks.

i) Any *-automorphism of $\underline{sp}(\mathcal{H}, J_Q; C_o)$ extends to a *-automorphism of $\underline{sl}(\mathcal{H}; C_o)$, and this in two different ways. Indeed, both φ_V and $X \longmapsto VJ_Q X^* J_Q V^*$ restrict to

$$\begin{cases} \underline{sp}(\mathcal{H}, J_Q; C_o) \longrightarrow \underline{sp}(\mathcal{H}, J_Q; C_o) \\ \qquad X \longmapsto VXV^* = VJ_Q X^* J_Q V^* \end{cases}$$

ii) In finite dimension, it is well known that all automorphisms of $C_n = \underline{sp}(n, \mathbb{C})$ are inner $(n > 3)$.

I.6. - Real forms

Definition 8. Let $\underline{g}$ be a complex involutive Lie algebra, with the involution denoted by $X \longmapsto X^*$. A conjugation of $\underline{g}$ is a semi-linear map $\sigma : \underline{g} \longrightarrow \underline{g}$ such that

$$\sigma^2(X) = X \quad \text{and} \quad \sigma(X^*) = (\sigma(X))^* \quad \text{for all } X \in \underline{g}$$

$$\sigma([X,Y]) = [\sigma(X), \sigma(Y)] \quad \text{for all } X,Y \in \underline{g} .$$

A real form of $\underline{g}$ is a real involutive Lie algebra $\underline{s}$ whose complexification is *-isomorphic to $\underline{g}$. The canonical conjugation of $\underline{g}$ is the map $X \longmapsto -X^*$ and its canonical real form, sometimes called the compact form of $\underline{g}$, is the real involutive Lie algebra $\underline{u} = \{X \in \underline{g} \mid X^* = -X\}$.

Standard results (Helgason [84] ch. III th. 7.1) show that a conjugation in the usual sense on a finite dimensional semi-simple Lie algebra is always a conjugation in the sense of definition 8 for a convenient c-involution on $\underline{g}$.

Let $\underline{g}$ and $\underline{u}$ be as above, let $\underline{s}$ be a real form of $\underline{g}$ and let $\varphi : \underline{u} \otimes_{\mathbb{R}} \mathbb{C} \longrightarrow \underline{s} \otimes_{\mathbb{R}} \mathbb{C}$ be a *-isomorphism. Let σ be the conjugation of $\underline{s} \otimes_{\mathbb{R}} \mathbb{C}$ defined by $\sigma(X + iY) = X - iY$ for all $X,Y \in \underline{s}$. Then the map $\tilde{\Gamma}_{\underline{s},\varphi} = \varphi^{-1}\sigma\varphi$ is a conjugation of $\underline{g}$ and the map

$$\Gamma_{\underline{s},\varphi} : \begin{cases} \underline{g} \longrightarrow \underline{g} \\ X \longmapsto \tilde{\Gamma}_{\underline{s},\varphi}(-X^*) \end{cases}$$

is an involutive *-automorphism of $\underline{g}$.

Conversely, let Γ be an involutive *-automorphism of $\underline{g}$. Then $\underline{s_\Gamma} = \{X \in \underline{g} \mid \Gamma(-X^*) = X\}$ is a real involutive Lie algebra and

$$\varphi_\Gamma : \begin{cases} \underline{g} \longrightarrow \underline{s_\Gamma} \otimes_{\mathbb{R}} \mathbb{C} \\ X \longmapsto \{\tfrac{1}{2}(X + \Gamma(-X^*))\} + i\,\{\tfrac{1}{2}i\,(X - \Gamma(-X^*))\} \end{cases}$$

is a *-isomorphism, so that $\underline{s_\Gamma}$ is a real form of $\underline{g}$.

In other words, there is a map from the set of all involutive $*$-automorphism of $\underline{g}$ to the set of $*$-isomorphism classes of real forms of $\underline{g}$, and this map is onto.

Let now Γ be an involutive $*$-automorphism of $\underline{g}$ and let Ψ be an arbitrary $*$-automorphism of $\underline{g}$. Then Γ and $\Psi^{-1}\Gamma\Psi$ define clearly $*$-isomorphic real forms of $\underline{g}$.

Conversely, let Γ_1 and Γ_2 be two $*$-automorphisms of $\underline{g}$ which define $*$-isomorphic real forms $\underline{s_1}$ and $\underline{s_2}$, let ψ be some $*$-isomorphism from $\underline{s_1}$ to $\underline{s_2}$ and let Ψ be the complexification of ψ. Then $\Gamma_2(-\Psi(Y)^*) = \Psi(Y)$ for all $Y \in \underline{s_1}$, so that $\Gamma_1(\Psi^{-1}\Gamma_2\Psi)(Y) = Y$ for all $Y \in \underline{s_1}$ hence for all $Y \in \underline{g}$. It follows that $\Gamma_1 = \Psi^{-1}\Gamma_2\Psi$.

These remarks boil down to the following proposition.

Proposition 11. Let $\underline{g}$ be a complex involutive Lie algebra. Then there is a bijective correspondence, as above, between

i) The set of conjugacy classes of involutive $*$-automorphisms of $\underline{g}$.

ii) The set of $*$-isomorphism classes of real forms of $\underline{g}$.

Alternative "proof". The Galois group $Z_2 = \mathrm{Gal}(\mathbb{C}/\mathbb{R})$ acts trivially on the group of $*$-automorphisms $\overset{*}{\mathrm{Aut}}(\underline{g}) = \overset{*}{\mathrm{Aut}}(\underline{u} \otimes_{\mathbb{R}} \mathbb{C})$, where $\underline{u}$ is the canonical real form of $\underline{g}$. According to general principles, the set of $*$-isomorphism classes of real forms of $\underline{g}$ is in bijection with the first cohomology $H^1(Z_2; \overset{*}{\mathrm{Aut}}(\underline{u} \otimes_{\mathbb{R}} \mathbb{C}))$. The rest of the proof follows directly from the definitions; see for example Serre [158], annexe to chap. VII and beginning of chap. X (mainly proposition 4). ∎

We now proceed to describe and classify the real forms of the classical complex Lie algebras of finite rank operators. The notations are close to those of Helgason [84]. Up to minor details, the method follows that of the three independent works by Balachandran [15], de la Harpe [76], [77] and Unsain [174], [175], devised for dealing with the L^*-version of the same problem. As in previous sections, $\mathcal{H}$ is an <u>infinite dimensional separable complex Hilbert space</u>. Generalisation to arbitrary dimensions is a question of notation only.

I.6A. - Real forms of $\underline{g} = \underline{sl}(\mathcal{H}, C_o)$

Type AI.

Let $J_{\mathbb{R}}$ be a conjugation of $\mathcal{H}$ and let σ_{AI} be the conjugation of $\underline{g}$ defined by $\sigma_{AI}(X) = J_{\mathbb{R}}XJ_{\mathbb{R}}$. The real form corresponding to σ_{AI} will be denoted by $\underline{sl}(\mathcal{H}_{\mathbb{R}}; C_o)$; it is indeed isomorphic to the involutive real Lie algebra consisting of all finite rank operators with trace zero in the real Hilbert space

$$\mathcal{H}_{\mathbb{R}} = \{x \in \mathcal{H} \mid J_{\mathbb{R}}x = x\} .$$

Type AII.

Let J_Q be an anticonjugation of $\mathcal{H}$ and let σ_{AII} be the conjugation of $\underline{g}$ defined by $\sigma_{AII}(X) = -J_QXJ_Q$. The corresponding real form is denoted by $\underline{su*}(\mathcal{H}; C_o)$.

Type AIII.

Let $r \in N \cup \{\infty\}$; let F_r be a r-dimensional subspace of $\mathcal{H}$ and let F^r be its orthogonal complement; if $r = \infty$, the space F^r is supposed to be infinite dimensional. Let I_r be the operator on $\mathcal{H}$ which coincides with minus the identity on F_r and with the identity on F^r. Let σ_{AIIIr} be the conjugation of $\underline{g}$ defined by $\sigma_{AIIIr}(X) = -I_r X^* I_r$. The corresponding real form is denoted by $\underline{su}(\mathcal{H}, r, \infty; C_o)$. When $r = 0$, it is the canonical real form of $\underline{g}$, which is called the compact form of $\underline{g}$, and denoted by $\underline{su}(\mathcal{H}; C_o)$.

Proposition 12A. Let $\underline{s}$ be a real form of $\underline{g}$. Then $\underline{s}$ is $*$-isomorphic to one of the c-involutive real Lie algebras defined above.

Proof.

Let φ be an involutive $*$-automorphism of $\underline{g}$. By proposition 10A, there exists a unitary operator V on $\mathcal{H}$ such that φ is either equal to φ_{VJ} or equal to φ_V.

Suppose first that $\varphi = \varphi_{VJ}$. As $\varphi^2 = id_{\underline{g}}$ the operator $VJ_{\mathbb{R}}VJ_{\mathbb{R}}$ commutes with any operator in $\underline{g}$; by Schur's lemma, it follows that there exists a complex number ζ of modulus one such that

(30) $$VJ_{\mathbb{R}}VJ_{\mathbb{R}} = \zeta id_{\mathcal{H}} .$$

Multiplying (30) on the right by $J_{\mathbb{R}}$ gives $\zeta J_{\mathbb{R}} = VJ_{\mathbb{R}}V$; multiplying (30) on the left by V^* and on the right by V gives $J_{\mathbb{R}}VJ_{\mathbb{R}}V = \zeta id_{\mathcal{H}}$. It follows that

$\bar{\zeta} id_{\mathcal{H}} = J_{\mathbb{R}}\zeta J_{\mathbb{R}} = J_{\mathbb{R}}VJ_{\mathbb{R}}V = \zeta id_{\mathcal{H}}$, so that $\zeta = \pm 1$.

If $\zeta = +1$, $VJ_{\mathbb{R}}$ is a conjugation of $\mathcal{H}$, so that the conjugation in $\underline{g}$ associated to φ is as in type AI above. If $\zeta = -1$, $VJ_{\mathbb{R}}$ is an anticonjugation of $\mathcal{H}$, so that the conjugation in $\underline{g}$ associated to φ is as in type AII above.

Suppose now that $\varphi = \varphi_V$. As $\varphi^2 = id_{\underline{g}}$, V^2 is a multiple of the identity of $\mathcal{H}$ and there is no loss of generality in supposing $V^2 = id_{\mathcal{H}}$ (see proposition 10A). Let F_r be the eigenspace of V corresponding to the eigenvalue -1 and let F^r be that corresponding to $+1$. Again, there is no loss of generality in supposing $\dim F_r \leq \dim F^r$. The conjugation in $\underline{g}$ associated to φ is then as one of those in type AIII described above. ■

Remarks.

i) Modulo an elementary checking, it follows from propositions 10A and 11 that the real forms described above are pairwise not *-isomorphic.

ii) Subsection I.6A could be repeated for $\underline{gl}(\mathcal{H}\,;\,C_o)$; see remark i) following proposition 10A.

I.6B. - Real forms of $\underline{g} = \underline{o}(\mathcal{H}, J_{\mathbb{R}};\, C_o)$

Type BDI.

Let $r \in N \cup \{\infty\}$ and let F_r , F^r and I_r be as above (subsection 6A, type AIII), with moreover $I_rJ_{\mathbb{R}} = J_{\mathbb{R}}I_r$. Let σ_{BDIr} be the conjugation of $\underline{g}$ defined by $\sigma_{BDIr}(X) = -I_rX^*I_r$. The corresponding real form is denoted by $\underline{o}(\mathcal{H}, r, \infty;\, C_o)$.

Alternatively, let $\mathcal{H}_{\mathbb{R}}$ be the real Hilbert space $\{x \in \mathcal{H} \mid J_{\mathbb{R}}x = x\}$ and let τ be the conjugation of $\underline{sl}(\mathcal{H}_{\mathbb{R}};\, C_o)$ defined formally as σ_{BDIr} . Then the involutive real Lie algebra $\underline{s} = \{X \in \underline{sl}(\mathcal{H}_{\mathbb{R}};\, C_o) \mid \tau(X) = X\}$ is *-isomorphic to $\underline{o}(\mathcal{H}, r, \infty;\, C_o)$; indeed, let T be the unitary operator on $\mathcal{H}$ which coincides with $(-i)$identity on F_r and with the identity on F^r ; then the map $X \longmapsto TXT^*$ is a *-automorphism of $\underline{sl}(\mathcal{H}\,;\,C_o)$ which sends $\underline{s}$ onto $\underline{o}(\mathcal{H}, r, \infty;\, C_o)$.

When $r = 0$, the conjugation σ_{BDIo} defines the canonical real form of $\underline{g}$, which is called the compact form of $\underline{g}$, and denoted by $\underline{o}(\mathcal{H}_{\mathbb{R}};\, C_o)$.

Type BDIII.

Let $J_{\mathbb{Q}}$ be an anticonjugation of $\mathcal{H}$ which commutes with $J_{\mathbb{R}}$ and let σ_{BDIII} be the conjugation of $\underline{g}$ defined by $\sigma_{BDIII}(X) = -J_{\mathbb{Q}}XJ_{\mathbb{Q}}$ $(= +J_{\mathbb{Q}}J_{\mathbb{R}}X^*J_{\mathbb{R}}J_{\mathbb{Q}})$. The corresponding real form is denoted by $\underline{o^*}(\mathcal{H}\,;\,C_o)$.

Proposition 12B. Let $\underline{s}$ be a real form of $\underline{g}$. Then $\underline{s}$ is *-isomorphic to one of the c-involutive real Lie algebras defined above.

Proof : as in the case of proposition 12A, the proof follows easily from proposition 10 and from Schur's lemma. ∎

I.6C. - Real forms of $\underline{g} = \underline{sp}(\mathcal{H}, J_{\mathbb{Q}};\, C_o)$

Type CI.

Let $J_{\mathbb{R}}$ be a conjugation of $\mathcal{H}$ which commutes with $J_{\mathbb{Q}}$ and let σ_{CI} be the conjugation of $\underline{g}$ defined by $\sigma_{CI}(X) = J_{\mathbb{R}}XJ_{\mathbb{R}}$. The corresponding real form is denoted by $\underline{sp}(\mathcal{H}, \mathbb{R};\, C_o)$.

Type CII.

Let $\mathcal{H}_I$ and $\mathcal{H}_{II}$ be two infinite dimensional subspaces of $\mathcal{H}$ such that $\mathcal{H} = \mathcal{H}_I \oplus \mathcal{H}_{II}$. Let $r \in N \cup \{\infty\}$; let F_{Ir} [resp. F_{IIr}] be a r-dimensional subspace of $\mathcal{H}_I$ [resp. $\mathcal{H}_{II}$] and let F_I^r [resp. F_{II}^r] be its orthogonal complement in $\mathcal{H}_I$ [resp. $\mathcal{H}_{II}$] ; if $r = \infty$, the spaces F_I^r and F_{II}^r are supposed to be infinite dimensional. The anticonjugation $J_{\mathbb{Q}}$ is supposed to map semi-isometrically E_{Ir} onto E_{IIr}, E_I^r onto E_{II}^r , and consequently E_{IIr} onto E_{Ir} and E_{II}^r onto E_I^r .

Let K_r be the operator on $\mathcal{H}$ which coincides with minus the identity on $E_{Ir} \oplus E_{IIr}$ and with the identity on $E_I^r \oplus E_{II}^r$ (in particular, K_r commutes with J_Q).

Let σ_{CIIr} be the conjugation of $\underline{g}$ defined by $\sigma_{CIIr}(X) = -K_r X^* K_r$. The corresponding real form is denoted by $\underline{sp}(\mathcal{H}, r, \infty; C_o)$. When $r = 0$, it is the canonical real form of $\underline{g}$, which is called the compact form of $\underline{g}$, and which is denoted by $\underline{sp}(\mathcal{H}_Q; C_o)$; it is indeed isomorphic to the involutive real Lie algebra consisting of all skew-adjoint finite rank operators in the quarternionic Hilbert space $\mathcal{H}_Q$ defined by $\mathcal{H}$ and J_Q.

Proposition 12C. Let $\underline{s}$ be a real form of $\underline{g}$. Then $\underline{s}$ is $*$-isomorphic to one of the c-involutive real Lie algebras defined above.

Proof : see proposition 12B. ■

I.6R. - Classical real Lie algebras of finite rank operators

Definition 9. A classical real Lie algebra of finite rank operators is one of the c-involutive simple real Lie algebras listed below. ($\mathcal{H}$ is assumed to be separable and infinite dimensional for simplicity of the notations, but the list could be made valid for a space of arbitrary dimension with minor modifications only.)

Type AI : $\underline{sl}(\mathcal{H}_{\mathbb{R}}; C_o)$.

AII : $\underline{su}^*(\mathcal{H} ; C_o)$.

AIII : $\underline{su}(\mathcal{H},r,\infty; C_o)$ where $r \in N \cup \{\infty\}$; the compact form of type A $\underline{su}(\mathcal{H} ; C_o)$ corresponds to $r = 0$.

Type BDI : $\underline{o}(\mathcal{H},r,\infty; C_o)$ where $r \in N \cup \{\infty\}$; the compact form of type BD $\underline{o}(\mathcal{H}_{\mathbb{R}}; C_o)$ corresponds to $r = 0$.

BDIII : $\underline{o}^*(\mathcal{H}; C_o)$.

Type CI : $\underline{sp}(\mathcal{H},\mathbb{R}; C_o)$.

CII : $\underline{sp}(\mathcal{H},r,\infty; C_o)$ where $r \in N \cup \{\infty\}$; the compact form of type C $\underline{sp}(\mathcal{H}_{\mathbb{Q}}; C_o)$ corresponds to $r = 0$.

Algebras with a complex structure : the algebras $\underline{sl}(\mathcal{H} ; C_o)$, $\underline{o}(\mathcal{H},J_{\mathbb{R}}; C_o)$ and $\underline{sp}(\mathcal{H},J_{\mathbb{Q}}; C_o)$ viewed as real Lie algebras.

These algebras are clearly all simple. The involution in each of them is defined by taking the adjoint of an operator in $\mathcal{H}$ (or in $\mathcal{H}_{\mathbb{R}}$, or in $\mathcal{H}_{\mathbb{Q}}$), and is a c-involution. The three algebras of the fourth group are the only ones which can be given a complex structure.

The Cartan subalgebras of the real Lie algebras of type A, B and C listed in definition 9 can be classified, using precisely the same

Appendix : conjugations and anticonjugations of a complex Hilbert space

Let V be a complex vector space of arbitrary dimension. The real vector space obtained by restricting the scalars will be denoted by $V^{\mathbb{R}}$. If σ is a map from V into itself, the map from $V^{\mathbb{R}}$ into itself induced by σ will be denoted by $\sigma^{\mathbb{R}}$. The map from $V^{\mathbb{R}}$ into itself induced by the multiplication by i in V will be denoted by J.

Lemma. Let V be a complex vector space and let σ be a semi-linear involution in V. Let $V_+ = \{x \in V^{\mathbb{R}} \mid \sigma^{\mathbb{R}}x = x\}$ and $V_- = \{x \in V^{\mathbb{R}} \mid \sigma^{\mathbb{R}}x = -x\}$. Then $V_- = JV_+$ and $V^{\mathbb{R}} = V_+ \oplus V_-$.

Proof. Define $\sigma_+ = \frac{1}{2}(\sigma^{\mathbb{R}} + 1)$ and $\sigma_- = \frac{1}{2}(\sigma^{\mathbb{R}} - 1)$. Clearly, $V_+ = \operatorname{im}\sigma_+$ and $V_- = \operatorname{im}\sigma_-$; as $x = \sigma_+(x) + \sigma_-(x)$ for all $x \in V$, the space V is the sum of V_+ and V_- ; as $V_+ \cap V_- = \{0\}$, the sum is direct. Finally, let $x \in V$; then $x \in J(V_+)$ if and only if $Jx \in V_+$, if and only if $Jx = \sigma_+(Jx)$, if and only if $Jx = -J\,\sigma_-(x)$, if and only if $x = \sigma_-(-x)$, if and only if $x \in V_-$; so that $JV_+ = V_-$. ∎

Let now $\mathcal{H}$ be a complex Hilbert space. For simplicity of the notations, $\mathcal{H}$ is supposed to be separable and infinite dimensional. A semi-linear operator $J_{\mathbb{R}}$ on $\mathcal{H}$ is a conjugation if $\langle J_{\mathbb{R}}x \mid J_{\mathbb{R}}y\rangle = \overline{\langle x \mid y\rangle}$ for all $x,y \in \mathcal{H}$ and if $J_{\mathbb{R}}^2 = 1$. The set of all fixed points of $J_{\mathbb{R}}$ supports a real Hilbert space $\mathcal{H}_{\mathbb{R}}$. In particular, there exists an orthonormal basis $e = (e_n)_{n \in N}$ in $\mathcal{H}$ such that $J_{\mathbb{R}}\left(\sum x_n e_n\right) = \sum \overline{x_n} e_n$ for all $x = \sum x_n e_n \in \mathcal{H}$; such a basis of $\mathcal{H}$ is said to be a $J_{\mathbb{R}}$-basis of type zero. Given $J_{\mathbb{R}}$, two J_R-basis of type zero are conjugated by an element of the full orthogonal group $O(\mathcal{H}_{\mathbb{R}})$. There exists as well an orthonormal basis $f = (f_n)_{n \in Z}$ such that $J_{\mathbb{R}} f_n = f_{-n}$ for all $n \in Z$. Indeed, let e be a J_R-basis of type zero; put $f_o = e_o$, and

$$f_{-n} = \frac{1}{\sqrt{2}}(e_{2n-1} + ie_{2n}) \ , \quad f_n = \frac{1}{\sqrt{2}}(e_{2n-1} - ie_{2n}) \quad \text{for all } n \in N^* \ .$$

A basis such as f is said to be a <u>$J_{\mathbb{R}}$-basis of type one.</u> Finally, there exists an orthonormal basis $g = (g_n)_{n \in Z^*}$ such that $J_{\mathbb{R}} g_n = g_{-n}$ for all $n \in Z^*$. Such a basis is said to be a <u>$J_{\mathbb{R}}$-basis of type two.</u>

A semi-linear operator $J_{\mathbb{Q}}$ on $\mathcal{H}$ is an <u>anticonjugation</u> if $\langle J_{\mathbb{Q}} x | J_{\mathbb{Q}} y \rangle = \overline{\langle x|y \rangle}$ for all $x,y \in \mathcal{H}$ and if $J_{\mathbb{Q}}^2 = -1$. The space furnished with the natural structure of quaternionic Hilbert space provided by $J_{\mathbb{Q}}$ will be denoted by $\mathcal{H}_{\mathbb{Q}}$. There exists an orthonormal basis $e = (e_n)_{n \in Z^*}$ in $\mathcal{H}$ such that

$$J_{\mathbb{Q}}\Big(\sum_{n \in N^*} x_{-n} e_{-n} + \sum_{n \in N^*} x_n e_n \Big) = \sum_{n \in N^*} \overline{x_{-n}} e_n - \sum_{n \in N^*} \overline{x_n} e_{-n}$$

for all $x = \sum_{n \in Z^*} x_n e_n \in \mathcal{H}$; such a basis is said to be a $J_{\mathbb{Q}}$-basis. Given $J_{\mathbb{Q}}$, two $J_{\mathbb{Q}}$-basis are conjugated by an element of the full symplectic group $Sp(\mathcal{H}_{\mathbb{Q}})$.

CHAPTER II .

CLASSICAL INVOLUTIVE BANACH-LIE ALGEBRAS AND GROUPS OF BOUNDED AND COMPACT OPERATORS

Let $g(\mathcal{H}; C_o)$ be a classical Lie algebra of finite rank operators. Its closure in some uniform crossnorm α will be denoted by $g(\mathcal{H}; C_\alpha)$ and its closure in the weak (or, for that matter, strong) topology by $g(\mathcal{H}; L)$. The aim of this Chapter is to study the Banach-Lie algebras obtained this way and the corresponding Banach-Lie groups of operators on $\mathcal{H}$. As in Chapter I, we usually assume the complex Hilbert space $\mathcal{H}$ to be separable and infinite dimensional in order to keep reasonably simple notations; but this is again a matter of convenience more than of necessity.

The results of this Chapter rely heavily on those of Chapter I. Numerous properties of operators in $L(\mathcal{H})$ have been used; though occasional, crucial help has been found in Herstein [86] and Johnson-Sinclair [92].

II.1.- Review of Banach-Lie groups and Banach-Lie algebras

The best reference for Banach-Lie groups is Lazard [107]; for shorter introductions, see Eells [54] section 3 and Lazard-Tits [108].

A Banach-Lie group is a (real or complex) Banach manifold G furnished with a group-structure which satisfies natural compatibility conditions. The tangent space $\underline{g}$ at the identity of G is then a Banach space and a Lie algebra with the multiplicity being jointly continuous; so that $\underline{g}$ is, by definition, a Banach-Lie algebra. The norm on $\underline{g}$ is defined by G up to equivalence only, and so involves an arbitrary choice; in all the following examples, the identity

$$\| [X,Y] \|_{\underline{g}} \leq 2 \| X \|_{\underline{g}} \| Y \|_{\underline{g}}$$

will be satisfied for all $X,Y \in \underline{g}$.

Standard properties of one parameter subgroups, the exponential map, adjoint representations and derivatives of smooth homomorphisms carry over from the finite dimensional to the Banach case. Similarly, if a Banach-Lie algebra is the algebra of any Banach-Lie group at all - in which case it is said to be enlargable - then it is the Lie algebra of a unique (up to isomorphism) connected and simply connected one.

A sub Banach-Lie group of a Banach-Lie group G is a subgroup H, furnished with a Banach manifold structure which makes the canonical injection of H into G smooth. It must be emphasized that H is not in general a submanifold of G ; indeed, the topology of H defined by its Banach manifold structure (referred to as its "own" topology below) might be strictly finer that its topology induced from G ; "submanifolds" of this type have already been defined by Chevalley [36], Chap. III §VI.

A <u>sub Banach-Lie algebra</u> of a Banach-Lie algebra $\underline{g}$ is a sub Lie algebra $\underline{h}$, furnished with a norm which makes it a Banach-Lie algebra and which makes the canonical injection of $\underline{h}$ into $\underline{g}$ continuous. The definitions are such that, given a Banach-Lie group G with Lie algebra $\underline{g}$, there is a bijective correspondence between sub Banach-Lie groups of G which are connected in their own topology and sub Banach-Lie algebras of $\underline{g}$.

Let G be a Banach-Lie group, let $\underline{g}$ be its Lie algebra, let H be a sub Banach-Lie groups of G and let $\underline{h}$ be the corresponding sub Banach-Lie algebra of $\underline{g}$. Then H is a submanifold in G if and only if

i) the own topology of H coincides with that induced by G

ii) $\underline{h}$ is closed and splits in $\underline{g}$.

If these conditions are fulfilled, H is sometimes said to be a <u>sub Banach-Lie group-manifold</u> of G. For example, any closed sub Banach-Lie group of G with a finite codimensional Lie algebra is a sub Banach-Lie group-manifold of G.

A Banach-Lie algebra is <u>simple</u> if it has no non-trivial closed ideal. It is <u>algebraically simple</u> if it has no non-trivial ideal at all. A Banach-Lie algebra $\underline{g}$ is provisionally said to be <u>semi-simple</u> if it has no closed abelian ideals and if its dimension is at least two (see the critic of this notion in an appendix to this Chapter). An <u>involutive Banach-Lie algebra</u> is a Banach-Lie algebra $\underline{g}$ furnished with an involution $X \mapsto X^*$ (def. I.1.) such that $\| X^* \| = \| X \|$ for all $X \in \underline{g}$.

Let E be a Banach space and let X be an operator on E. Let $\{X\}^+$ be the closed associative subalgebra of L(E) generated by X. Then X is said to be <u>semi-simple</u> if $\{X\}^+$ does not contain any non-zero quasi-nilpotent element. An element X in a Banach-Lie algebra $\underline{g}$ is <u>semi-simple</u> if ad(X) is a semi-simple operator on $\underline{g}$. A <u>c-involutive Banach-Lie algebra</u> is an involutive Banach-Lie algebra in which any normal element is semi-simple. The definition of <u>Cartan subalgebras</u> given in section I.3 carries over to the Banach case without change.

II.4

Unless otherwise stated, homomorphisms [resp. *-homomorphisms] between Banach-Lie algebras [resp. involutive Banach-Lie algebras] are not supposed to be necessarily continuous. Hence a slightly different terminology as in Balachandran [12], [14], or as in [76]. Definition I.8 (about conjugations and real forms) and proposition I.11 (about the cohomological formulation of the problem of classifying real forms) carry over to the Banach case with the minor evident modifications. For example, a <u>conjugation</u> of an involutive Banach-Lie algebra $\underline{g}$ is a map $\sigma : \underline{g} \to \underline{g}$ such that the map $X \longmapsto \sigma(-X^*)$ is an involutive *-automorphism of $\underline{g}$; the corresponding <u>real form</u> is given by $\{X \in \underline{g} | \sigma(X) = X\}$.

II.2.- Classical complex Banach-Lie algebras of bounded operators : definitions and ideal-structure

In the theory of associative involutive Banach algebras, one of the first examples usually given is the C^*-algebra $L(\mathcal{H}_{\mathbb{K}})$ of continuous operators in a Hilbert space $\mathcal{H}_{\mathbb{K}}$ over $\mathbb{K}$. In the case $\mathcal{H}_{\mathbb{K}}$ is separable and infinite dimensional, it is well-known that any associative proper ideal of $L(\mathcal{H}_{\mathbb{K}})$ contains the ideal $C_o(\mathcal{H}_{\mathbb{K}})$ of finite rank operators and is contained in the ideal $C(\mathcal{H}_{\mathbb{K}})$ of compact operators (see for example Schatten [151] chap. I th. 11); in particular, $C(\mathcal{H}_{\mathbb{K}})$ is the only non-trivial closed ideal of $L(\mathcal{H}_{\mathbb{K}})$ and the algebra $L(\mathcal{H}_{\mathbb{K}})/C(\mathcal{H}_{\mathbb{K}})$ is algebraically simple. This can be summed up graphically in the following picture:

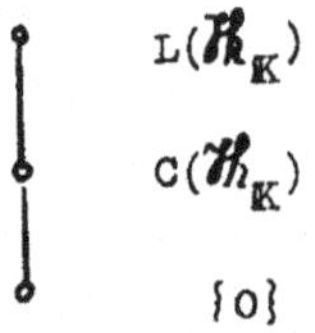

The aim of this section is to point out analogous facts in the Lie case; the results are those of section 5 in [126].

Let $\mathcal{H}$ be a complex Hilbert space, say <u>separable and infinite dimensional</u>. The associative involutive Banach algebra $L(\mathcal{H})$ defines an involutive Banach-Lie algebra which will be denoted by $\underline{gl}(\mathcal{H}; L)$, or sometimes simply by $\underline{gl}(\mathcal{H})$. The following subalgebras are clearly all ideals in $\underline{gl}(\mathcal{H})$:

$\mathbb{C}\, id_{\mathcal{H}}$ = center of $\underline{gl}(\mathcal{H})$.

$\underline{sl}(\mathcal{H}; C_o)$ = subalgebra of finite rank operators with 0-trace.

$\underline{gl}(\mathcal{H}; C_o)$ = subalgebra of all finite rank operators.

$\underline{gl}(\mathcal{H}; C)$ = subalgebra of all compact operators.

$gl(\mathcal{H}; \mathbb{C}id_{\mathcal{H}} + C) = \{X \in \underline{gl}(\mathcal{H}) \mid X = \lambda id_{\mathcal{H}} + Y$ for some $\lambda \in \mathbb{C}$, and some $Y \in C(\mathcal{H})\}$.

<u>Proposition 1A</u>. Let $\underline{a}$ be a non-trivial ideal of $\underline{gl}(\mathcal{H})$ which is distinct from the center. Then

$$\underline{sl}\,(\mathcal{H};\, C_o) \subset \underline{a} \subset \underline{gl}(\mathcal{H};\, \mathbb{C}id_{\mathcal{H}} + C).$$

In particular, the only non-trivial closed ideals of $\underline{gl}(\mathcal{H})$ are the center, $\underline{gl}(\mathcal{H};\, C)$ and $\underline{gl}(\mathcal{H};\, \mathbb{C}id_{\mathcal{H}} + C)$; the Banach-Lie algebra $\underline{gl}(\mathcal{H};\, L) / \underline{gl}(\mathcal{H};\, \mathbb{C}id_{\mathcal{H}} + C)$ is algebraically simple. The picturial summing up for closed ideals is:

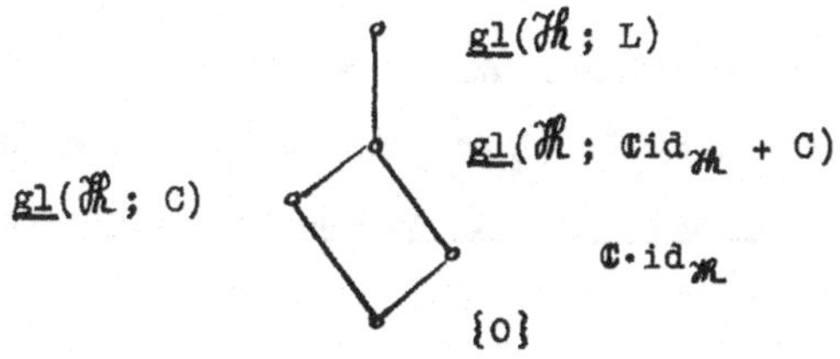

<u>Proof</u>.

<u>Step one</u> : $\underline{sl}(\mathcal{H};\, C_o) \subset \underline{a}$.

Choose $X \in \underline{a}$, X not a multiple of the identity. Either X is normal, or $[X,X^*]$ is not a multiple of the identity (Putnam [134] chap. I th. 1.2.1). Hence X can be supposed normal without loss of generality. By Schur's lemma, there exists $Y \in \underline{sl}(\mathcal{H};\, C_o)$ such that $[X,Y] \neq 0$; hence $\underline{a} \cap \underline{sl}(\mathcal{H};\, C_o)$ is a non-zero ideal of $\underline{sl}(\mathcal{H};\, C_o)$. Step one follows then from proposition I.1A.

<u>Step two</u> : the derived ideal of $\underline{gl}(\mathcal{H})$ is $\underline{gl}(\mathcal{H})$ itself.

This is a well-known result due to Halmos; see for example Dieudonné [45] chap. XV §11 exercise 23.

<u>Step three</u> : the center of the simple associative algebra $L(\mathcal{H}) / C(\mathcal{H})$ consists exactly of the multiples of the identity. See Calkin [30]th. 2.9.

<u>Step four</u> : let $\mathcal{R}$ be a simple associative algebra, with center $\mathcal{Z}$, over a field of characteristic not 2; let $\underline{a}$ be a Lie ideal in $\mathcal{R}$;

then either $\underline{a} \subset \mathcal{Z}$, or $[\mathcal{A},\mathcal{A}] \subset \underline{a}$. See Herstein [86], th. 4; or for a cheaper but here clearly sufficient result, Bourbaki [25] §1 exercise 7.

<u>Step five</u>. As a consequence of steps two to four, a Lie ideal of $L(\mathcal{H})/C(\mathcal{H})$ is either trivial or equal to the center. Hence any Lie ideal of $L(\mathcal{H})$ is contained in $\underline{gl}(\mathcal{H}; \mathbb{C}id_{\mathcal{H}} + C)$. As $\underline{sl}(\mathcal{H}; C_o)$ is dense in $\underline{gl}(\mathcal{H}; C)$ (see the appendix, proposition II.19. iv), and as $\underline{gl}(\mathcal{H}; C)$ is of codimension one in $\underline{gl}(\mathcal{H}; \mathbb{C}id_{\mathcal{H}} + C)$, the proposition follows. ∎

Let now $J_{\mathbb{R}}$ be a conjugation of $\mathcal{H}$. The involutive antiautomorphism of $L(\mathcal{H})$ defined by $X \mapsto J_{\mathbb{R}}X^*J_{\mathbb{R}}$ is again denoted by $\varphi_{\mathbb{R}}$ (see section I.1). The <u>orthogonal complex Banach-Lie algebra</u> corresponding to $\underline{gl}(\mathcal{H})$ is

$$\underline{o}(\mathcal{H}, J_{\mathbb{R}}; L) = \{X \in \underline{gl}(\mathcal{H}; L) \mid \varphi_{\mathbb{R}}(X) = -X\}.$$

and will sometimes be simply denoted by $\underline{o}(\mathcal{H}, J_{\mathbb{R}})$. Its center is reduced to zero and the following subalgebras are clarly both ideals in $\underline{o}(\mathcal{H}, J_{\mathbb{R}})$:

$\underline{o}(\mathcal{H}, J_{\mathbb{R}}; C_o)$ = subalgebra of finite rank operators.

$\underline{o}(\mathcal{H}, J_{\mathbb{R}}; C)$ = subalgebra of compact operators.

<u>Proposition 1B</u>. Let $\underline{a}$ be a non-trivial ideal of $\underline{o}(\mathcal{H}, J_{\mathbb{R}})$. Then $\underline{o}(\mathcal{H}, J_{\mathbb{R}}; C_o) \subset \underline{a} \subset \underline{o}(\mathcal{H}, J_{\mathbb{R}}; C)$.

In particular, the only non-trivial closed ideal of $\underline{o}(\mathcal{H}, J_{\mathbb{R}})$ is $\underline{o}(\mathcal{H}, J_{\mathbb{R}}; C)$; the Banach-Lie algebra $\underline{o}(\mathcal{H}, J_{\mathbb{R}}; L)/\underline{o}(\mathcal{H}, J_{\mathbb{R}}; C)$ is algebraically simple. The pictural summing up for closed ideals is:

$\underline{o}(\mathcal{H}, J_{\mathbb{R}}; L)$
$\underline{o}(\mathcal{H}, J_{\mathbb{R}}; C)$
$\{0\}$

Proof.

Step one : $\underline{o}(\mathcal{M},J_{\mathbb{R}};\ C_o) \subset \underline{a}$. This step is proved as the corresponding one for proposition 1A; note that $\underline{o}(\mathcal{M},J_{\mathbb{R}})$ does not contain the identity.

Step two ; let $\mathcal{B}$ be a set of generators in an associative algebra $\mathcal{R}$; then $[\mathcal{B},\mathcal{R}] = [\mathcal{R},\mathcal{R}]$.

For each $n \in N^*$, let $\mathcal{B}^n$ be the vector space generated in $\mathcal{R}$ by products of n elements of $\mathcal{B}$; we first show by induction on n that $[\mathcal{B}^n,\mathcal{R}] \subset [\mathcal{B},\mathcal{R}]$ for all $n \in N^*$. It is trivial when n = 1; it follows in the general case from the induction hypothesis via the computation :

$$[X_1X_2\ldots X_n\ ,\ Y] = X_1(X_2\ldots X_nY) - (X_2\ldots X_nY)X_1 + X_2\ldots X_n(YX_1) -$$
$$-(YX_1)X_2\ldots X_n \in [\mathcal{B},\mathcal{R}] + [\mathcal{B}^{n-1},\mathcal{R}] \subset [\mathcal{B},\mathcal{R}]$$

for all $X_1, \ldots, X_n \in \mathcal{B}$, for all $Y \in \mathcal{R}$. Now, to say that $\mathcal{B}$ is a set of generators in $\mathcal{R}$ is to say that $\mathcal{R} = \sum_{n\in N^*} \mathcal{B}^n$; hence :

$$[\mathcal{R},\mathcal{R}] \subset \sum_{n\in N^*} [\mathcal{B}^n,\mathcal{R}] \subset [\mathcal{B},\mathcal{R}] \subset [\mathcal{R},\mathcal{R}]$$ and step two follows.

Step three : $\underline{o}(\mathcal{M},J_{\mathbb{R}})$ is a set of generators in the associative algebra $L(\mathcal{M})$; compare with Herstein [86]th. 8.

For the proof of this step, let $\mathcal{R}$ denote for short $L(\mathcal{M})$, let $\underline{o}$ denote $\underline{o}(\mathcal{M},J_{\mathbb{R}};\ L)$, let $\mathcal{A}$ be the associative algebra generated by $\underline{o}$ in $\mathcal{R}$, and let $\underline{m} = \{X \in L(\mathcal{M}) \mid \varphi_{\mathbb{R}}(X) = X\}$, so that $\mathcal{R} = \underline{o} \oplus \underline{m}$.

Let $A,B \in \underline{o}$, and let $P \in \underline{m}$. Thus $PA + AP \in \underline{o}$, and so $(PA + AP)B \in \mathcal{A}$. Similarly $(PB + BP)A \in \mathcal{A}$. By subtraction : $P[A,B] + APB - BPA \in \mathcal{A}$. However, $APB - BPA = APB - \varphi_{\mathbb{R}}(APB)$ is in $\underline{o}$, and so is in $\mathcal{A}$. Consequently $P[A,B] \in \mathcal{A}$, or otherwise said, $\underline{m}[A,B] \subset \mathcal{A}$ for all $A,B \in \underline{o}$. But clearly $\underline{o}[A,B] \subset \mathcal{A}$ for all $A,B \in \underline{o}$, so that $\mathcal{R}[A,B] = (\underline{o} \oplus \underline{m})[A,B] \subset \mathcal{A}$ for all $A,B \in \underline{o}$. Similarly, $[C,D]\mathcal{R} \subset \mathcal{A}$ for all $C,D \in \underline{o}$. Hence $\mathcal{R}[A,B]\ [C,D]\mathcal{R} \subset \mathcal{A}$ for all $A,B,C,D \in \underline{o}$.

$\mathcal{R}[A,B]\ [C,D]\mathcal{R}$ is clearly a two-sided ideal in $\mathcal{R}$. Hence, either it is inside $C(\mathcal{H})$, or it is the whole of $\mathcal{R} = L(\mathcal{H})$. By a convenient choice of A,B,C,D, this ideal can clearly be forced in the second case, so that $\mathcal{R} \subset \mathcal{A}$, which proves step three.

Step four : the derived ideal of $\underline{o}(\mathcal{H},J_{\mathbb{R}})$ is $\underline{o}(\mathcal{H},J_{\mathbb{R}})$ itself.

Let $\underline{o}$ and $\underline{m}$ be as in step three. By step two in the proof of proposition 1A and by steps two and three above :

$\underline{gl}(\mathcal{H}) = [\underline{gl}(\mathcal{H}), \underline{gl}(\mathcal{H})] = [\underline{o}, \underline{o}\oplus\underline{m}] \subset [\underline{o},\underline{o}] + [\underline{o},\underline{m}]$. As $[\underline{o},\underline{o}] \subset \underline{o}$ and as $[\underline{o},\underline{m}] \subset \underline{m}$, the conclusion follows.

Step five : let $\mathcal{R}$ be a simple associative algebra, with center $\mathcal{Z}$, over a field of characteristic not 2, and such that the dimension of $\mathcal{R}$ over $\mathcal{Z}$ is larger than 16 ; let φ be an involutive antiautomorphism of $\mathcal{R}$ and let $\underline{a}$ be a Lie ideal in $\underline{o} = \{X \in \mathcal{R} \mid \varphi(X) = -X\}$; then either $\underline{a} \subset \mathcal{Z}$, or $[\underline{o},\underline{o}] \subset \underline{a}$. See Herstein [86], th. 9.

Step six : As a consequence of step three in the proof of proposition 1A and of steps four and five above, the only Lie ideals of $\underline{o}(\mathcal{H},J_{\mathbb{R}};\, L)/\underline{o}(\mathcal{H},J_{\mathbb{R}};\, C)$ are the trivial ones; hence any Lie ideal of $\underline{o}(\mathcal{H},J_{\mathbb{R}})$ is contained in $\underline{o}(\mathcal{H},J_{\mathbb{R}};\, C)$. As $\underline{o}(\mathcal{H},J_{\mathbb{R}};\, C_o)$ is dense in $\underline{o}(\mathcal{H},J_{\mathbb{R}};\, C)$, the proposition follows. ∎

Let now $J_{\mathbb{Q}}$ be an anticonjugation of $\mathcal{H}$. The involutive antiautomorphisms of $L(\mathcal{H})$ defined by $X \mapsto -J_{\mathbb{Q}}X^*J_{\mathbb{Q}}$ is again denoted by $\varphi_{\mathbb{Q}}$. The <u>symplectic complex Banach-Lie algebra</u> corresponding to $\underline{gl}(\mathcal{H})$ is

$$\underline{sp}(\mathcal{H},J_{\mathbb{Q}};\ L) = \{X \in \underline{gl}(\mathcal{H};\, L) \mid \varphi_{\mathbb{Q}}(X) = -X\},$$

and will sometimes be simply denoted by $\underline{sp}(\mathcal{H},J_{\mathbb{Q}})$. Its center is reduced to zero and the following subalgebras are clearly both ideals in $\underline{sp}(\mathcal{H},J_{\mathbb{Q}})$:

$\underline{sp}(\mathcal{H},J_{\mathbb{Q}};\, C_o)$ = subalgebra of finite rank operators.

$\underline{sp}(\mathcal{H},J_{\mathbb{Q}};\, C)$ = subalgebra of compact operators.

<u>Proposition 1C</u>. Let $\underline{a}$ be a non-trivial ideal of $\underline{sp}(\mathcal{H},J_{\mathbb{Q}})$. Then

$$\underline{sp}(\mathcal{H},J_{\mathbb{Q}};\, C_o) \subset \underline{a} \subset \underline{sp}(\mathcal{H},J_{\mathbb{Q}};\, C).$$

In particular, the only non-trivial closed ideal of $\underline{sp}(\mathcal{H},J_{\mathbb{Q}})$ is $\underline{sp}(\mathcal{H},J_{\mathbb{Q}};C)$; the Banach-Lie algebra $\underline{sp}(\mathcal{H},J_{\mathbb{Q}};L)/\underline{sp}(\mathcal{H},J_{\mathbb{Q}};C)$ is algebraically simple. The pictorial summing up for closed ideals is similar to that given in proposition 1B.

<u>Proof</u> : as for proposition 1B. ∎

<u>Definition 1</u>. <u>A classical complex Banach-Lie algebra of bounded operators</u> is one of the involutive Banach-Lie algebras

$\underline{gl}(\mathcal{H};L)$ $\qquad$ $\underline{o}(\mathcal{H},J_{\mathbb{R}};L)$ $\qquad$ $\underline{sp}(\mathcal{H},J_{\mathbb{Q}};L)$

as defined above, where $\mathcal{H}$ is a complex Hilbert space and where $J_{\mathbb{R}}$ [resp. $J_{\mathbb{Q}}$] is some conjugation [resp. anticonjugation] of $\mathcal{H}$.

II.3.- Classical complex Banach-Lie algebras of bounded operators : derivations, automorphisms and real forms

The information collected in Chapter I about the classical Lie algebras of finite rank operators can be used in the study of the classical Banach-Lie algebras of bounded operators. The three propositions which follow are respectively corollaries of sections I.2, I.5 and I.6; the analogue of section I.3 on Cartan subalgebras is left to the reader; the consideration of roots (section I.4) does not seem to me to be of any further help for the knowledge of the algebras of the present Chapter.

In this section, $\mathcal{H}$ is an infinite dimensional complex Hilbert space.

Proposition 2.

A.- Let Δ be a derivation of $\underline{gl}(\mathcal{H})$. Then there exists a unique operator D (up to addition of a scalar multiple of the identity) in $\underline{gl}(\mathcal{H})$ such that $\Delta = ad(D)$.

B.- Let Δ be a derivation of $\underline{o}(\mathcal{H},J_{\mathbb{R}})$. Then there exists a unique operator D in $\underline{o}(\mathcal{H},J_{\mathbb{R}})$ such that $\Delta = ad(D)$.

C.- Let Δ be a derivation of $\underline{sp}(\mathcal{H},J_{\mathbb{Q}})$. Then there exists a unique operator D in $\underline{sp}(\mathcal{H},J_{\mathbb{Q}})$ such that $\Delta = ad(D)$.

Proof (case A). Let Δ be a derivation of $\underline{gl}(\mathcal{H})$.

Step one : definition of D and δ.

As $\underline{sl}(\mathcal{H};C_o)$ is an ideal in $\underline{gl}(\mathcal{H})$ which is equal to its own derived ideal, it is globally invariant by Δ. According to proposition I.2A, there exists an endomorphism $D \in Lin(\mathcal{H})$ such that $\Delta(Y) = [D,Y]$ for all $Y \in \underline{sl}(\mathcal{H};C_o)$. Let then δ be the function defined by

$$\begin{cases} \underline{gl}(\mathcal{H}) \longrightarrow Lin(\mathcal{H}) \\ X \longmapsto \Delta(X) - [D,X] \end{cases}.$$

Step two : δ is zero.

Choose $X \in \underline{gl}(\mathcal{H})$; for every $Y \in \underline{sl}(\mathcal{H}; C_o)$:

$$\Delta([X,Y]) = \Big[D,[X,Y]\Big] = \Big[\Delta(X),Y\Big] + \Big[X,\ \Delta(Y)\Big] =$$

$$= \Big[[D,X],Y\Big] + \Big[\delta(X),Y\Big] + \Big[X,[D,Y]\Big].$$

It follows from Jacobi identity that $\delta(X)$ commutes with all finite rank operators having zero trace, i.e. that $\delta(X)$ is multiple of the identity, say $\delta(X) = \varepsilon(X)id_{\mathcal{H}}$. As Δ is linear, ε defines a linear form or $\underline{gl}(\mathcal{H})$. From the derivation rule for Δ, it follows easily that ε vanishes on commutators. As $\underline{gl}(\mathcal{H})$ is equal to its derived ideal, it follows that ε vanishes, hence so does δ.

Step three : Δ is continuous.

Steps one and two show that Δ is already a derivation of the associative algebra $L(\mathcal{H})$. The end of the proof follows then from standard properties of derivations in C^*- and von Neumann algebras; see Dixmier [46] chap III §9. The fact that the continuity of Δ implies that of D is left to the reader.

Modifications for cases B and C. Let Δ be a derivation of $\underline{o}(\mathcal{H},J_{\mathbb{R}})$. Steps one and two can be repeated almost without change, so that $\Delta(X) = [D,X]$ for all $X \in \underline{o}(\mathcal{H},J_{\mathbb{R}})$, where D is an ad hoc operator on $\mathcal{H}$. It follows that Δ can be extended to a derivation of $\underline{gl}(\mathcal{H})$, whence the continuity of D, whence that of Δ. This can be repeated for $\underline{sp}(\mathcal{H},J_{\mathbb{Q}})$. ∎

Proposition 2 can be rephrased as follows.

Corollary. Let $\underline{g}$ be a classical complex Banach-Lie algebra of bounded operators. Then any derivation of $\underline{g}$ is continuous and inner.

In particular, any Lie derivation of the C^*-algebra $L(\mathcal{H})$ is already a derivation in the associative sense.

Proposition 3.

A.- Let φ be a *-automorphism of $\underline{gl}(\mathcal{H})$ and let $J_{\mathbb{R}}$ be a fixed conjugation on $\mathcal{H}$. Then there exists a unitary operator $V \in U(\mathcal{H})$ such that

$$\text{either } \varphi = \varphi_{VJ} : \begin{cases} \underline{gl}(\mathcal{H}) & \longrightarrow & \underline{gl}(\mathcal{H}) \\ X & \longmapsto & -VJ_{\mathbb{R}}X^*J_{\mathbb{R}}V^* \end{cases}$$

$$\text{or } \varphi = \varphi_V : \begin{cases} \underline{gl}(\mathcal{H}) & \longrightarrow & \underline{gl}(\mathcal{H}) \\ X & \longmapsto & VXV^* \end{cases}$$

The two cases exclude each other.

The operator V is uniquely determined by φ, up to multiplication by a complex number of modulus one. Otherwise said, the sequence

$$\{1\} \longrightarrow U(1) \xrightarrow{j} \tilde{U}(\mathcal{H}) \xrightarrow{\pi} \overset{*}{\mathrm{Aut}}(\underline{gl}(\mathcal{H})) \longrightarrow \{1\}$$

is exact (notations as in proposition I.10A).

Similarly and more briefly in the two other cases :

B.- The sequence $\{1\} \rightarrow Z_2 \longrightarrow O(\mathcal{H}_{\mathbb{R}}) \longrightarrow \overset{*}{\mathrm{Aut}}(\underline{o}(\mathcal{H},J_{\mathbb{R}})) \rightarrow \{1\}$ is exact.

C.- The sequence $\{1\} \rightarrow Z_2 \longrightarrow Sp(\mathcal{H}_{\mathbb{Q}}) \longrightarrow \overset{*}{\mathrm{Aut}}(\underline{sp}(\mathcal{H},J_{\mathbb{Q}})) \rightarrow \{1\}$ is exact.

Proof (case A). Let φ be a *-automorphism of $\underline{gl}(\mathcal{H})$.

Step one. As $\underline{sl}(\mathcal{H};C_o)$ is absolutely minimal among the ideals of $\underline{gl}(\mathcal{H})$ having dimension strictly bigger than one (proposition 1), it is globally invariant by φ. According to proposition I.10A, there exists $V \in U(\mathcal{H})$ such that the restriction of φ to $\underline{sl}(\mathcal{H};C_o)$ is either φ_{VJ} or φ_V.

Step two. Suppose first that $\varphi(Y) = \varphi_V(Y)$ for all $Y \in \underline{sl}(\mathcal{H};C_o)$. Choose $X \in \underline{gl}(\mathcal{H})$; for any $Y \in \underline{sl}(\mathcal{H};C_o)$:
$\varphi([X,Y]) = V[X,Y]V^* = [VXV^*,VYV^*] = [\varphi(X),\varphi(Y)] = [\varphi(X),VYV^*]$. It follows from Schur's lemma that $\varphi(X) - VXV^*$ is a multiple of the identity, say $\varphi(X) - VXV^* = \varepsilon(X)id_{\mathcal{H}}$. As φ is linear, ε defines a

linear form on $\underline{gl}(\mathcal{H})$. From the automorphism rule for φ, it follows easily that ε vanishes on commutators. As $\underline{gl}(\mathcal{H})$ is equal to its derived ideal, it follows that ε vanishes; hence $\varphi = \varphi_V$.

Step three. Suppose now that $\varphi(Y) = \varphi_{VJ}(Y)$ for all $Y \in \underline{sl}(\mathcal{H}; C_o)$. The same argument as in step two shows that $\varphi(X) = \varphi_{VJ}(X)$ for all $X \in \underline{gl}(\mathcal{H})$. ∎

Proposition 3 can be rephrased as follows.

Corollary. Let $\underline{g}$ be a classical complex Banach-Lie algebra of bounded operators. Then any *-automorphism of $\underline{g}$ is isometric and inner.

In particular, any Lie *-automorphism of the C^*-algebra $L(\mathcal{H})$ is already either a *-automorphism or the negative of a *-antiautomorphism in the associative sense.

Remarks.

i) The case A of proposition 2 follows as well from general facts about derivations of von Neumann algebras (Dixmier [46] chap. III §9 th. 1) and about Lie derivations of primitive rings (Martindale III [117] th. 2).

ii) The case A of proposition 3 follows as well from general facts about *-automorphisms of von Neumann algebras (Dixmier [46] chap. III § 3, n° 2) and about Lie automorphisms of prime rings (Martindale [118] th. 13). With the help of other methods, it is possible to show that all (not necessarily *-) automorphisms of $\underline{gl}(\mathcal{H})$ are inner : see for example Arnold [3] th. 4.

iii) I do not know of any published results which would be the analogues of those in [117] and [118] for rings with involution, and of which propositions 2B, 2C, 3B and 3C would be typical examples. However, these propositions are again corollaries of particular results due to Rickart [140].

iv) The word "inner" in the corollary above has a slightly broader sense than in the associative case : in proposition 3A indeed, the group $\tilde{U}(\mathcal{H})$ appears, not only its subgroup $U(\mathcal{H})$.

The considerations of section I.6 can now be repeated with almost no change for classical Banach-Lie algebras of bounded operators. Definition I.8 and proposition I.11 have evident analogues in the present context, and so have each of the conjugations and real forms described in section I.6. For example, the conjugation of $\underline{gl}(\mathcal{H})$ defined by $X \longmapsto J_{\mathbb{R}}XJ_{\mathbb{R}}$ is again denoted by σ_{AI} and the corresponding real form is $\underline{gl}(\mathcal{H}_{\mathbb{R}}; L) = \{X \in \underline{gl}(\mathcal{H}; L) \mid \sigma_{AI}(X) = X\}$. The other involutive Banach-Lie algebras appearing in the following definition can be expressed similarly.

Definition 2. A classical real Banach-Lie algebra of bounded operators is one of the involutive Banach-Lie algebras listed below. ($\mathcal{H}$ is assumed to be separable and infinite dimensional for simplicity of the notations, but the list could be made valid for a space of arbitrary dimension with minor modifications only.)

Type AI : $\underline{gl}(\mathcal{H}_{\mathbb{R}}; L)$.

AII : $\underline{u^*}(\mathcal{H}; L)$.

AIII : $\underline{u}(\mathcal{H}, r, \infty; L)$ where $r \in N \cup \{\infty\}$; the compact form of type A $\underline{u}(\mathcal{H}; L)$ corresponds to $r = 0$.

Type BDI : $\underline{o}(\mathcal{H}, r, \infty; L)$ where $r \in N \cup \{\infty\}$; the compact form of type BD $\underline{o}(\mathcal{H}_{\mathbb{R}}; L)$ corresponds to $r = 0$.

BDIII : $\underline{o^*}(\mathcal{H}; L)$.

Type CI : $\underline{sp}(\mathcal{H}, \mathbb{R}; L)$.

CII : $\underline{sp}(\mathcal{H}, r, \infty; L)$ where $r \in N \cup \{\infty\}$; the compact form of type C $\underline{sp}(\mathcal{H}_{\mathbb{Q}}; L)$ corresponds to $r = 0$.

There are analogues to proposition 1 and 2 for each classical real Banach-Lie algebra of bounded operators. Both their statements and their proofs are left to the reader.

<u>Proposition 4</u>. Let <u>g</u> be a classical complex Banach-Lie algebra of bounded operators and let <u>s</u> be a real form of <u>g</u>. Then <u>s</u> is *-isomorphic to a classical real Banach-Lie algebra of bounded operators, of one of the types A, B and C, as described in definition 2.

<u>Proof</u> : see propositions I.12 and II.3. ■

II.4.- Classical Banach-Lie groups of bounded operators

Each of the Banach-Lie algebras described in the two preceding sections is the Lie algebra of a Banach-Lie group of operators.

For example $\underline{gl}(\mathcal{H}; L)$ is the Lie algebra of the general linear group $GL(\mathcal{H}; L)$, or for short $GL(\mathcal{H})$, of all invertible operators on $\mathcal{H}$. The subgroup of $GL(\mathcal{H})$ consisting of those elements which leave invariant the bilinear form

$$\begin{cases} \mathcal{H} \times \mathcal{H} \longrightarrow \mathbb{C} \\ (x,y) \longrightarrow \langle x | J_{\mathbb{R}} y \rangle \end{cases}$$

is denoted by $O(\mathcal{H}, J_{\mathbb{R}}; L)$ or by $O(\mathcal{H}, J_{\mathbb{R}})$; it is a sub Banach-Lie group-manifold of $GL(\mathcal{H})$ and its Lie algebra is clearly $\underline{o}(\mathcal{H}, J_{\mathbb{R}})$. Similarly, the subgroup of $GL(\mathcal{H})$ consisting of those elements which leave invariant the bilinear form

$$\begin{cases} \mathcal{H} \times \mathcal{H} \longrightarrow \mathbb{C} \\ (x,y) \longrightarrow \langle x | J_{\mathbb{Q}} y \rangle \end{cases}$$

is denoted by $Sp(\mathcal{H}, J_{\mathbb{Q}}; L)$ or by $Sp(\mathcal{H}, J_{\mathbb{Q}})$; it is again a sub Banach-Lie group-manifold of $GL(\mathcal{H})$ and its Lie algebra is clearly $\underline{sp}(\mathcal{H}, J_{\mathbb{Q}})$.

<u>Definition 3</u>. A <u>classical complex Banach-Lie group of bounded operators</u> is one of the Banach-Lie groups $GL(\mathcal{H}; L)$, $O(\mathcal{H}, J_{\mathbb{R}}; L)$ and $Sp(\mathcal{H}, J_{\mathbb{Q}}; L)$ as defined above, where $\mathcal{H}$ is a complex Hilbert space and where $J_{\mathbb{R}}$ [resp. $J_{\mathbb{Q}}$] is some conjugation [resp. anticonjugation] of $\mathcal{H}$.

The definition and the list of the <u>classical real Banach-Lie groups of bounded operators</u> is left to the reader (see definition 2 section II.3).

In the rest of this section, we review some of the standard facts known about the groups introduced above. The set of selfadjoints [resp. positive] operators in $L(\mathcal{H})$ is denoted by $\mathrm{Sym}(\mathcal{H})$ [resp. $\mathrm{Pos}(\mathcal{H})$].

Proposition 5. Let $G(\mathcal{H})$ be a classical real or complex Banach-Lie group of bounded operators on $\mathcal{H}$ and let $\underline{g}(\mathcal{H})$ be its Lie algebra.

i) The exponential map $\underline{g}(\mathcal{H}) \longrightarrow G(\mathcal{H})$ is given by the traditional power series.

ii) The exponential map provides an analytic morphism, which is a local isomorphism

$$\underline{g}(\mathcal{H}) \cap \underline{u}(\mathcal{H}) \xrightarrow{\quad \exp \quad} G(\mathcal{H}) \cap U(\mathcal{H}).$$

iii) The exponential map provides an analysic isomorphism

$$\underline{g}(\mathcal{H}) \cap \mathrm{Sym}(\mathcal{H}) \xrightarrow{\quad \exp \quad} G(\mathcal{H}) \cap \mathrm{Pos}(\mathcal{H}).$$

iv) The polar decomposition provides an analytic isomorphism

$$G(\mathcal{H}) \longrightarrow \Big(G(\mathcal{H}) \cap U(\mathcal{H})\Big) \times \Big(G(\mathcal{H}) \cap \mathrm{Pos}(\mathcal{H})\Big)$$

Proof : for i), see Lazard [107] prop. 10.6 and remarque 21.6; for ii) to iv), see Lang [103] chap. 7 prop. 3, 5 and 6 when $G(\mathcal{H}) = GL(\mathcal{H})$; the other cases can be proved the same way. ∎

Proposition 6.

A.- The exponential map $\underline{u}(\mathcal{H}; L) \longrightarrow U(\mathcal{H}; L)$ is onto.

B.- The image of the exponential map $\underline{o}(\mathcal{H}_{\mathbb{R}}; L) \longrightarrow O(\mathcal{H}_{\mathbb{R}}; L)$ is the set of those operators such that the multiplicity with which -1 appears in their point spectrum is either finite and even (possibly zero) or infinite.

C.- The exponential map $\underline{sp}(\mathcal{H}_{\mathbb{Q}}; L) \longrightarrow Sp(\mathcal{H}_{\mathbb{Q}}; L)$ is onto.

Proof: in Putnam and Winter [136], section 7 to 11; their argument, for $\mathbb{K} = \mathbb{R}$, can be easily adapted (and simplified) for the cases $\mathbb{K} = \mathbb{C}$ and $\mathbb{K} = \mathbb{Q}$. ∎

<u>Proposition 7</u>. Let $\mathcal{H}$ be infinite dimensional. Then the real Banach-Lie groups $U(\mathcal{H})$, $O(\mathcal{H}_{\mathbb{R}})$ and $Sp(\mathcal{H}_{\mathbb{Q}})$ are contractible. Similarly, the complex Banach-Lie groups $GL(\mathcal{H})$, $O(\mathcal{H}, J_{\mathbb{R}})$ and $Sp(\mathcal{H}, J_{\mathbb{Q}})$ are contractible.

<u>Proof</u> : due to Kuiper [101], [89]; for a recent discussion on this result, see Mityagini [123]. ∎

The homotopy type of any other classical Banach-Lie group of operators on $\mathcal{H}$ follows trivially from proposition 7, via proposition II.5. iv.

It is often possible to translate in the context of the classical Banach-Lie groups of bounded operators properties of their Lie algebras as those seen earlier in this chapter.

For example, Brown and Pearcy have shown that any element in $GL(\mathcal{H}; L)$ is a product of two commutators ([134] section I.4).

There are detailed studies, due to Kadison, of the normal closed subgroups of $GL(\mathcal{H})$, $U(\mathcal{H})$, and of the general linear and unitary groups in the other types of factors [93], [94], [95]. According to Rosenberg ([144], p. 279), some of these results have been further improved by Kaplansky (unpublished). Similar problems have been dealt with by Sunouchy [171].

It seems a sound conjecture, and probably even an easy one (though tedious) to check, that analogues of these various results about the derived groups and the normal subgroups of $GL(\mathcal{H})$ and $U(\mathcal{H})$ still hold for the other classical Banach-Lie groups of bounded operators.

II.5.- Classical Banach-Lie algebras of compact operators

In this section, $\mathcal{H}$ is a complex Hilbert space, of infinite dimension if not otherwise stated. Properties of the ideals $C_p(\mathcal{H})$ are recalled in an appendix to this chapter.

Let $p \in \bar{\mathbb{R}}$, $1 \leq p \leq \infty$. The associative involutive Banach algebra $C_p(\mathcal{H})$ defines an involutive Banach-Lie algebra which will be denoted by $\underline{gl}(\mathcal{H}; C_p)$, or by $\underline{gl}(\mathcal{H}; C)$ when $p = \infty$. When $p = 1$, the closure of its derived ideal is the involutive Banach-Lie algebra consisting of nuclear operators with trace zero, and will be denoted by $\underline{sl}(\mathcal{H}; C_1)$.

Let $J_{\mathbb{R}}$ be a conjugation of $\mathcal{H}$. The involutive antiautomorphism of $C_p(\mathcal{H})$ defined by $X \longmapsto J_{\mathbb{R}}X^*J_{\mathbb{R}}$ is again denoted by $\varphi_{\mathbb{R}}$ (see sections I.1 and II.2). The <u>orthogonal complex Banach-Lie algebra</u> corresponding to $\underline{gl}(\mathcal{H}; C_p)$ is

$$\underline{o}(\mathcal{H}, J_{\mathbb{R}}; C_p) = \{X \in \underline{gl}(\mathcal{H}; C_p) \mid \varphi_{\mathbb{R}}(X) = -X\}.$$

Let $J_{\mathbb{Q}}$ be an anticonjugation of $\mathcal{H}$. The involutive anti-automorphism of $C_p(\mathcal{H})$ defined by $X \mapsto -J_{\mathbb{Q}}X^*J_{\mathbb{Q}}$ is again denoted by $\varphi_{\mathbb{Q}}$. The <u>symplectic complex Banach-Lie algebra</u> corresponding to $\underline{gl}(\mathcal{H}; C_p)$ is

$$\underline{sp}(\mathcal{H}, J_{\mathbb{Q}}; C_p) = \{X \in \underline{gl}(\mathcal{H}; C_p) \mid \varphi_{\mathbb{Q}}(X) = -X\}.$$

<u>Definition 4</u>. A <u>classical complex Banach-Lie algebra of compact operators</u> is one of the involutive Banach-Lie algebras

$\underline{gl}(\mathcal{H}; C_p)$ $\quad$ $\underline{sl}(\mathcal{H}; C_1)$ $\quad$ $\underline{o}(\mathcal{H}, J_{\mathbb{R}}; C_p)$ $\quad$ $\underline{sp}(\mathcal{H}, J_{\mathbb{Q}}; C_p)$

as defined above, where $\mathcal{H}$ is a complex Hilbert space, where $J_{\mathbb{R}}$ [resp. $J_{\mathbb{Q}}$] is some conjugation [resp. anticonjugation] of $\mathcal{H}$, and where p is in $\bar{\mathbb{R}}$, with $1 \leq p \leq \infty$.

Many properties of these algebras follow straightforwardly from those of the Lie algebras studied so far. The end of this section will list some of them; proofs, being evident, will be non-existent or very sketchy.

<u>Proposition 8</u>. Let $\underline{g}(\mathcal{H}; C_p)$ be a classical complex Banach-Lie algebra of compact operator on $\mathcal{H}$. Then the involution defined by $X \mapsto X^*$ is a c-involution; and any nonzero ideal of $\underline{g}(\mathcal{H}; C_p)$ contains the corresponding classical complex Banach-Lie algebra of finite rank operators. In particular, $\underline{g}(\mathcal{H}; C_p)$ is a (topologically) simple c-involutive Banach-Lie algebra (except $\underline{gl}(\mathcal{H}; C_1)$ which is not simple).

<u>Proof</u> : see propositions I.1 and II.1. ∎

<u>Definition 5</u>. Let $\underline{g}(\mathcal{H}; C_p)$ be a classical complex Banach-Lie algebra of compact operators on $\mathcal{H}$. A derivation Δ of $\underline{g}(\mathcal{H}; C_p)$ is said to be <u>spatial</u> if there exists an operator D in the corresponding classical complex Banach-Lie algebra of bounded operators such that $\Delta(X) = [D,X]$ for all $X \in \underline{g}(\mathcal{H}; C_p)$. A *-automorphism φ of $\underline{g}(\mathcal{H}; C_p)$ is said to be <u>spatial</u> if there exists an operator V in the corresponding classical complex Banach-Lie group of bounded operators such that one of the following holds:

either $\varphi(X) = -VJ_{\mathbb{R}}X^*J_{\mathbb{R}}V^*$ for all $X \in \underline{g}(\mathcal{H}; C_p)$, for some conjugation $J_{\mathbb{R}}$ on $\mathcal{H}$,

or $\varphi(X) = VXV^*$ for all $X \in \underline{g}(\mathcal{H}; C_p)$.

<u>Proposition 9</u>. Any derivation of a classical complex Banach-Lie algebra of compact operators on $\mathcal{H}$ is continuous and spatial.

<u>Proof</u> : see propositions I.2, II.2 and corollary; the reference given in the step three of the proof of propositions II.2 must now be

replaced by Johnson and Sinclair [92]. The situation described by proposition II.2 and II.9 is clearly reminiscent of that known for simple C^*-algebras [146]. ∎

<u>Proposition 10</u>. The Cartan subalgebras of a classical complex Banach-Lie algebra of compact operators are described exactly as in section I.3 : those of $\underline{gl}(\mathcal{H}; C_p)$ and of $\underline{sp}(\mathcal{H}, J_{\mathbb{Q}}; C_p)$ are conjugated by the group of *-automorphisms; those of $\underline{o}(\mathcal{H}, J_{\mathbb{R}}; C_p)$ split into two conjugation-classes under the action of the group of *-automorphism.

<u>Proof</u> : see proposition I.3. ∎ Note that a Cartan subalgebra of a classical complex Banach-Lie algebra of compact operators on $\mathcal{H}$ contains a dense set of regular elements if and only if $\mathcal{H}$ is separable [9]. In all cases, such a subalgebra is equal to its normalizer.

The notions of <u>root</u>, <u>root vector</u>, and <u>type</u> of a Cartan subalgebra of $\underline{o}(\mathcal{H}, J_{\mathbb{R}}; C_p)$, are defined in the evident way.

<u>Proposition 11</u>. Let $\underline{g}$ be a classical complex Banach-Lie algebra of compact operators and let $\underline{h}$ be a Cartan subalgebra of $\underline{g}$. Then the roots of $\underline{g}$ with respect to $\underline{h}$ are given by the formulae (I.6) to (I.19).

<u>Proof</u> : section I.4. ∎

<u>Proposition 12</u>. Any *-automorphism of a classical complex Banach-Lie algebra of compact operators on $\mathcal{H}$ is isometric and spatial.

<u>Proof</u> : propositions I.10 and II.3. ∎

The considerations of section I.6 on real forms can again be repeated with almost no change. The notations should be by now obvious enough for us to give without further comment the following definition.

<u>Definition 6</u>. A <u>classical real Banach-Lie algebra of compact operators</u> is one of the c-involutive Banach-Lie algebras listed below (the list is given for $\mathcal{H}$ separable and infinite dimensional; $p \in \bar{\mathbb{R}}$ with $1 \leq p \leq \infty$).

<u>Type AI</u> : $\underline{gl}(\mathcal{H}_{\mathbb{R}};\ C_p)$ and $\underline{sl}(\mathcal{H}_{\mathbb{R}};\ C_1)$.

AII : $\underline{u}^*(\mathcal{H};\ C_p)$ and $\underline{su}^*(\mathcal{H};\ C_1)$.

AIII : $\underline{u}(\mathcal{H},r,\infty;\ C_p)$ and $\underline{su}(\mathcal{H},r,\infty;\ C_1)$ where $r \in N \cup\{\infty\}$; the compact forms of type A $\underline{u}(\mathcal{H};\ C_p)$ and $\underline{su}(\mathcal{H};\ C_1)$ correspond to $r = 0$.

<u>Type BDI</u> : $\underline{o}(\mathcal{H},r,\infty;\ C_p)$ where $r \in N \cup\{\infty\}$; the compact form of type B $\underline{o}(\mathcal{H}_{\mathbb{R}};\ C_p)$ corresponds to $r = 0$.

BDIII : $\underline{o}^*(\mathcal{H};\ C_p)$.

<u>Type CI</u> : $\underline{sp}(\mathcal{H},\mathbb{R};\ C_p)$.

CII : $\underline{sp}(\mathcal{H},r,\infty;\ C_p)$ where $r \in N \cup\{\infty\}$; the compact form of type C $\underline{sp}(\mathcal{H}_{\mathbb{Q}};\ C_p)$ corresponds to $r = 0$.

<u>Algebras with a complex structure</u> : the algebras $\underline{gl}(\mathcal{H};\ C_p)$, $\underline{sl}(\mathcal{H};\ C_1)$, $\underline{o}(\mathcal{H},J_{\mathbb{R}};\ C_p)$ and $\underline{sp}(\mathcal{H},J_{\mathbb{Q}};\ C_p)$ viewed as real Lie algebras.

<u>Proposition 13</u>. Let $\underline{g}$ be a classical complex Banach-Lie algebra of compact operators and let $\underline{s}$ be a real form of $\underline{g}$. Then $\underline{s}$ is *-isomorphic to a classical real Banach-Lie algebra of compact operators, of one of the types A, B and C, as described in definition 6.

<u>Proof</u> : see section I.6 and proposition II.4. ∎

We end this section with two remarks.

Pearcy and Topping [131] have shown that the derived ideal of $\underline{gl}(\mathcal{H}; C)$ is the whole algebra, and that the derived ideal of $\underline{gl}(\mathcal{H}; C_{2p})$ is exactly $\underline{gl}(\mathcal{H}; C_p)$ for all $p \in \mathbb{R}$ such that $p > 1$; it is not known if the derived ideal of $\underline{gl}(\mathcal{H}; C_2)$ is the whole of $\underline{sl}(\mathcal{H}; C_1)$. Similar results can be conjectured (and are probably easy to prove) for the corresponding orthogonal and symplectic complex Lie algebras.

There is a sort of duality among the Banach-Lie algebras introduced in this Chapter which might suggest a generalization of the notion of L^*-algebra. More precisely, let $\underline{a}$ and $\underline{b}$ be two sub involutive Banach-Lie algebras of $\underline{gl}(\mathcal{H}; L)$; then $\underline{b}$ is said to be <u>dual to $\underline{a}$ in $\underline{g}$</u> if there exists a sesquilinear pairing $\langle\!\langle \mid \rangle\!\rangle : \underline{a} \times \underline{b} \longrightarrow \mathbb{C}$ and if the following conditions are satisfied :

i) $[\underline{a},\underline{b}]$ is an ideal both in $\underline{a}$ and in $\underline{b}$;

ii) $\langle\!\langle [A,X] \mid Y \rangle\!\rangle = \langle\!\langle X \mid [A^*,Y] \rangle\!\rangle$ for all $A,X \in \underline{a}$, for all $Y \in \underline{b}$;

iii) $\langle\!\langle X \mid [B,Y] \rangle\!\rangle = \langle\!\langle [B^*,X] \mid Y \rangle\!\rangle$ for all $X \in \underline{a}$, for all $B,Y \in \underline{b}$;

iv) the map
$$\begin{cases} \underline{b} \longrightarrow (\underline{a})^{\text{dual}} \\ B \longmapsto \begin{pmatrix} X & \underline{a} \\ \downarrow & \downarrow \\ \langle\!\langle X \mid B^* \rangle\!\rangle & \mathbb{C} \end{pmatrix} \end{cases}$$
defines an isometry of the Banach space $\underline{b}$ onto the strong dual of the Banach space $\underline{a}$.

Examples of such a situation are given by :

$$\begin{aligned} \underline{a} &= \underline{gl}(\mathcal{H}; C) & \underline{b} &= \underline{gl}(\mathcal{H}; C_1) \\ &= \underline{gl}(\mathcal{H}; C_p) & &= \underline{gl}(\mathcal{H}; C_q) \\ &\quad (p \in \mathbb{R},\ 1 < p < \infty) & &\quad (q \in \mathbb{R},\ \tfrac{1}{p} + \tfrac{1}{q} = 1) \\ &= \underline{gl}(\mathcal{H}; C_1) & &= \underline{gl}(\mathcal{H}; L) \end{aligned}$$

and by the corresponding orthogonal complex and symplectic complex pairs; in all cases, the pairing is given by $\langle\!\langle X \mid Y \rangle\!\rangle = \text{trace}(XY^*)$. The "duality" carries over to Cartan subalgebras, roots, ... The project might be suggested to try and extend (at least part of) Schue's results [153], [154] to "pairs of involutive Banach-Lie algebras", according to some definition tailored from the above examples. (Schue's L^*-case is given by $p = q = 2$.)

II.6.- Classical Banach-Lie groups of compact operators

This section describes Banach-Lie groups which correspond to the Banach-Lie algebras of section II.5 in the same way as the groups described in section II.4 correspond to the algebras of section II.2 and II.3.

Let $p \in \bar{\mathbb{R}}$, $1 \leq p \leq \infty$ and let $GL(\mathcal{H}; C_p)$ be the group of those invertible operators on $\mathcal{H}$ which can be written as $id_{\mathcal{H}} + X$, where X is in $C_p(\mathcal{H})$. As there is a distinguished bijection between $GL(\mathcal{H}; C_p)$ and an open subset in $C_p(\mathcal{H})$, the group is in fact a sub Banach-Lie group of $GL(\mathcal{H}; L)$ - not a sub group-manifold (II.3) - and its Lie algebra is $\underline{gl}(\mathcal{H}; C_p)$. When $p = 1$, $SL(\mathcal{H}; C_1)$ is the subgroup of $GL(\mathcal{H}; C_1)$ consisting of those operators having determinant +1 ; it is again a Banach-Lie group, with Lie algebra $\underline{sl}(\mathcal{H}; C_1)$.

Similarly, one defines for each $p \in \bar{\mathbb{R}}$, $1 \leq p \leq \infty$, the Banach-Lie groups $O(\mathcal{H}, J_{\mathbb{R}}; C_p)$ whose connected component is denoted by $O^+(\mathcal{H}, J_{\mathbb{R}}; C_p)$ and whose Lie algebra is $\underline{o}(\mathcal{H}, J_{\mathbb{R}}; C_p)$; $Sp(\mathcal{H}, J_{\mathbb{Q}}; C_p)$ whose Lie algebra is $\underline{sp}(\mathcal{H}, J_{\mathbb{Q}}; C_p)$; and the groups of operators corresponding to the Lie algebras of definition 6. These groups are respectively called the <u>classical complex Banach-Lie groups of compact operators</u> and the <u>classical real Banach-Lie groups of compact operators</u>.

Unless otherwise stated, we will always consider the <u>connected components</u> of these groups. When explicitely written down, the connected component of a group will be affected by a subscript + ; for example, $O^+(\mathcal{H}_{\mathbb{R}}; C_p)$ is of index 2 in $O(\mathcal{H}_{\mathbb{R}}; C_p)$. (Hence, if $\mathcal{H}$ was finite dimensional, $O^+(\mathcal{H}_{\mathbb{R}})$ would denote the Lie group classically written $SO(\mathcal{H}_{\mathbb{R}})$).

The properties of these groups corresponding to propositions 5 to 7 of section II.4 will now be listed. The space of self-adjoint operators in $C_p(\mathcal{H})$ is denoted by $Sym(\mathcal{H}; C_p)$. The set of those positive operators on $\mathcal{H}$ which can be written as $id_{\mathcal{H}} + X$ with $X \in C_p(\mathcal{H})$ is denoted by $Pos(\mathcal{H}; C_p)$.

Proposition 14. Let $G(\mathcal{H}; C_p)$ be a classical real or complex Banach-Lie group of compact operators on $\mathcal{H}$ and let $\underline{g}(\mathcal{H}; C_p)$ be its Lie algebra.

i) The exponential map $\underline{g}(\mathcal{H}; C_p) \longrightarrow G(\mathcal{H}; C_p)$ is given by the traditional power series.

ii) The exponential map provides an analytic morphism, which is a local isomorphism

$$\underline{g}(\mathcal{H}; C_p) \cap \underline{u}(\mathcal{H}; C_p) \xrightarrow{\exp} G(\mathcal{H}; C_p) \cap U(\mathcal{H}; C_p).$$

iii) The exponential map provides an analytic isomorphism

$$\underline{g}(\mathcal{H}; C_p) \cap Sym(\mathcal{H}; C_p) \xrightarrow{\exp} G(\mathcal{H}; C_p) \cap Pos(\mathcal{H}; C_p).$$

iv) The polar decomposition provides an analytic isomorphism from $G(\mathcal{H}; C_p)$ to

$$\left(G(\mathcal{H}; C_p) \cap U(\mathcal{H}; C_p)\right) \times \left(G(\mathcal{H}; C_p) \cap Pos(\mathcal{H}; C_p)\right).$$

Proof : see proposition II.5. ■

Proposition 15

A.- The exponential map $\underline{u}(\mathcal{H}; C_p) \longrightarrow U(\mathcal{H}; C_p)$ is onto.

B.- The image of the exponential map $\underline{o}(\mathcal{H}_{\mathbb{R}}; C_p) \to O(\mathcal{H}_{\mathbb{R}}; C_p)$ is the set of those operators such that the multiplicity with which -1 appears in their spectrum is (finite and) even (possibly zero).

C.- The exponential map $\underline{sp}(\mathcal{H}_{\mathbb{Q}}; C_p) \to Sp(\mathcal{H}_{\mathbb{Q}}; C_p)$ is onto.

Proof : see proposition II.6; case B is easier than the original result of Putnam and Wintner insofar as the spectral theorem for compact operators is good enough; a detailed proof was written up in [75]. ■

The statement corresponding to proposition II.7 is a corollary of a much more general theorem, due to Geba [64] and extending former results by Elworthy [61], [62], Palais [129] and Svarc [172]. We first give the result proved, though not stated in this generality, by by Geba.

Let E be an infinite dimensional Banach space over $\mathbb{K} \in \{\mathbb{R}, \mathbb{C}, \mathbb{Q}\}$. Let $\mathcal{P}(E)$ be a subspace of L(E), furnished with a norm (but not necessarily complete), and enjoying the following properties :

i) If $X \in \mathcal{P}(E)$, $id_{\mathcal{H}} + X$ is a Fredholm operator on E ; equivalently, $\mathcal{P}(E)$ is contained in the set of Riesz operators on E (see Schechter [152]).

ii) If $X \in \mathcal{P}(E)$, then $X + C_0(E) \subset \mathcal{P}(E)$; in particular, $C_0(E) \subset \mathcal{P}(E)$.

iii) The multiplication $C_0(E) \times \mathcal{P}(E) \longrightarrow \mathcal{P}(E)$ is continuous, where $C_0(E)$ is furnished with the uniform norm and $\mathcal{P}(E)$ with its own norm.

Let then $GL(E;\mathcal{P})$ be the subset of GL(E) consisting of all those invertible operators on E which can be written as $id_{\mathcal{H}} + X$, with $X \in \mathcal{P}(E)$; let $GL(E;\mathcal{P})$ be endowed with the topology inherited from the norm given on $\mathcal{P}(E)$. Note that if $\mathcal{P}(E)$ is an ideal in L(E), then $GL(E;\mathcal{P})$ is a group; if $\mathcal{P}(E)$ is a normed algebra, the multiplication is continuous in $GL(E;\mathcal{P})$; if $\mathcal{P}(E)$ is a Banach algebra, $GL(E;\mathcal{P})$ is a Banach-Lie group. Numerous examples of such spaces $\mathcal{P}(E)$ can be found for example in Pietsch [132], [133].

<u>Proposition 16</u>. Let E and $GL(E;\mathcal{P})$ be as above. Then $GL(E;\mathcal{P})$ is homotopically equivalent to the stable general linear group $GL(\infty, \mathbb{K})$.

<u>Proof</u> : see Geba [64]. ∎

Note that, if $e = (e_n)_{n \in N}$ is an orthonormal basis of the Hilbert space $\mathcal{H}_{\mathbb{K}} = E$, if $GL(n, \mathbb{K})$ is identified via e to the evident subgroup of $GL(\mathcal{H}_{\mathbb{K}}; \mathcal{C})$, and if $GL(\infty, \mathbb{K})$ is the inductive limit of the $GL(n, \mathbb{K})$'s, then the inclusion $GL(\infty, \mathbb{K}) \to GL(\mathcal{H}_{\mathbb{K}}; \mathcal{C})$ is itself a homotopy equivalence. The inclusions of the corollary are constructed the same way.

<u>Corollary</u>. Let $\mathcal{H}$ be infinite dimensional, separable and complex; let $p \in \bar{\mathbb{R}}$, $1 \leq p \leq \infty$. Then the homotopy types of the classical Banach-Lie groups of compact operators are given by the following table, where all arrows are homotopy equivalences.

<u>Complex groups</u>.

<u>Type A</u> : $S^1 \times SU(\infty) \sim GL(\infty, \mathbb{C}) \longrightarrow GL(\mathcal{H}; C_p)$.
$SU(\infty) \longrightarrow SL(\infty, \mathbb{C}) \longrightarrow SL(\mathcal{H}; C_1)$.

<u>Type B</u> : $SO(\infty) \longrightarrow SO(\infty, \mathbb{C}) \longrightarrow O^+(\mathcal{H}, J_{\mathbb{R}}; C_p)$.

<u>Type C</u> : $Sp(\infty) \longrightarrow Sp(\infty, \mathbb{C}) \longrightarrow Sp(\mathcal{H}, J_{\mathbb{Q}}; C_p)$.

<u>Real groups</u>.

<u>Type AI</u> : $SO(\infty) \longrightarrow GL^+(\infty, \mathbb{R}) \longrightarrow GL^+(\mathcal{H}_{\mathbb{R}}; C_p)$.
$SO(\infty) \longrightarrow SL(\infty, \mathbb{R}) \longrightarrow SL(\mathcal{H}_{\mathbb{R}}; C_1)$.

AII : $S^1 \times Sp(\infty) \sim U^*(\infty) \longrightarrow U^*(\mathcal{H}; C_p)$.
$Sp(\infty) \longrightarrow SU^*(\infty) \longrightarrow SU^*(\mathcal{H}; C_1)$.

AIII : if $r = 0$ $\quad S^1 \times SU(\infty) \sim U(\mathcal{H}; C_p)$.
$SU(\infty) \longrightarrow SU(\mathcal{H}; C_1)$.

if $r \in N$ $\quad S^1 \times SU(r) \times S^1 \times SU(\infty) \sim U(\mathcal{H}, r, \infty; C_p)$
$SU(r) \times S^1 \times SU(\infty) \sim SU(\mathcal{H}, r, \infty; C_1)$.

if $r = \infty$ $\quad S^1 \times SU(\infty) \times S^1 \times SU(\infty) \sim U(\mathcal{H}, \infty, \infty; C_p)$.
$S(U(\infty) \times U(\infty)) \sim SU(\mathcal{H}, \infty, \infty; C_1)$.

<u>Type BDI</u> : if $r = 0$ $\quad SO(\infty) \longrightarrow O^+(\mathcal{H}_{\mathbb{R}}; C_p)$.
if $r \in N$ $\quad SO(r) \times SO(\infty) \longrightarrow O^+(\mathcal{H}, r, \infty; C_p)$.

if $r = \infty$ $\quad$ $SO(\infty) \times SO(\infty) \longrightarrow O^+(\mathcal{H},\infty,\infty; C_p)$.

BDIII $\quad$: $\quad U(\infty) \longrightarrow SO^*(\infty) \longrightarrow O^*(\mathcal{H}; C_p)$.

Type CI $\quad$: $S^1 \times SU(\infty) \;\sim\; Sp(\infty, \mathbb{R}) \longrightarrow Sp(\mathcal{H},\mathbb{R}; C_p)$.

CII $\quad$: if $r = 0$ $\quad$ $Sp(\infty) \longrightarrow Sp(\mathcal{H}_{\mathbb{Q}}; C_p)$.

$\qquad\qquad$ if $r \in \mathbb{N}$ $\quad$ $Sp(r) \times Sp(\infty) \longrightarrow Sp(\mathcal{H},r,\infty; C_p)$.

$\qquad\qquad$ if $r = \infty$ $\quad$ $Sp(\infty) \times Sp(\infty) \longrightarrow Sp(\mathcal{H},\infty,\infty; C_p)$.

Proof : via propositions II.14.iv) and II.16. ■

$$* \quad * \quad * \quad * \quad *$$

The "classical Banach-Lie groups of operators" considered so far are by no means the only examples which appear naturally as sub Banach-Lie groups of the general linear group of a Hilbert (or Banach) space. Indeed, the following remarks show that two standard constructions in the theory of Banach spaces provide rich sources of examples.

Groups tied to extention problems

Let F and E be two Banach spaces over $\mathbb{K}$ ($\mathbb{K} = \mathbb{R}$ or $\mathbb{K} = \mathbb{C}$) and let $F \xrightarrow{i} E$ be a dense continuous linear injection. Let $\underline{g}$ be the sub Lie algebra of $L(F)$ consisting of those operators X on F which factor through continuous operators from E to F:

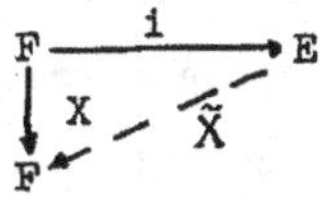

$$
\begin{array}{ccc}
F & \xrightarrow{\;i\;} & E \\
\downarrow{\scriptstyle X} & \underset{\tilde{X}}{\swarrow} & \\
F & &
\end{array}
$$

Then $\underline{g}$ is a Banach-Lie algebra for the norm $|X| = \|\tilde{X}\|_{L(E,F)}$. (One can suppose the norm of i to be bounded by 1 for the identity page II.2 to hold.) Let G be the group of those invertible operators on F of the form $\mathrm{id}_F + X$, where X factors through E. Then G has a

Banach-Lie group structure with Lie algebra $\underline{g}$. If i is, for example, compact, then G is a sub Banach-Lie group of the Fredholm group of F; other restrictions on i are of interest (p-summability, radonifying maps). Note that, in general, the ideal of finite rank operators $C_o(F)$ does not belong to $\underline{g}$, so that the condition ii) of page II.28 is not satisfied. However :

Conjecture : If i is compact, then the group G defined above is homotopically equivalent to $GL(K^\infty)$.

To prove the conjecture, one has probably to modify Geba's argument (see proposition II.16) for the situations where pairs of Banach spaces are involved. Another method of attacking the problem would be to try and apply Cerf's result ([34] chap. III §1) to the inclusion of G into the Fredholm group of F.

Another procedure to manufacture normed Lie algebras is to consider extention problems summarized by the diagram

$$\begin{array}{ccc} F & \xrightarrow{\ i\ } & E \\ \downarrow & & \vdots \\ F & \xrightarrow{\ i\ } & E \end{array}$$

Groups tied to lifting problems

Let $E \xrightarrow{\ j\ } F$ be a dense continuous linear injection between two real or complex Banach spaces and let $\underline{g}$ be the sub Banach-Lie algebra of L(F) consisting of those operators X on F which factor through continuous operators from F to E :

$$\begin{array}{ccc} & \tilde{\tilde{X}} & F \\ & \swarrow & \downarrow X \\ E & \xrightarrow{\ j\ } & F \end{array}$$

Let G be the group of those invertible operators on F of the form $id_F + X$, where X factors through E. Then G has a Banach-Lie group structure with Lie algebra $\underline{g}$. The above conjecture can be repeated in this case when j is compact.

Another procedure to manufacture normed Lie algebras is again to consider lifting problems summarized by the diagram.

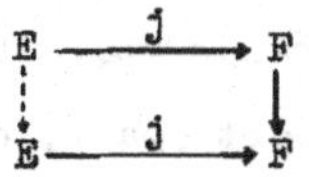

Mixed Problems

We will give one example only of a construction involving both extension and lifting: Let ($\mathcal{H} \xrightarrow{i} E$) be an Abstract Wiener Space (see Gross [68]). The Wiener group defined in the introduction (Section 1.1) can be thought as resulting of a two-stages construction according to the diagram ($\mathcal{H}^* = \mathcal{H}$)

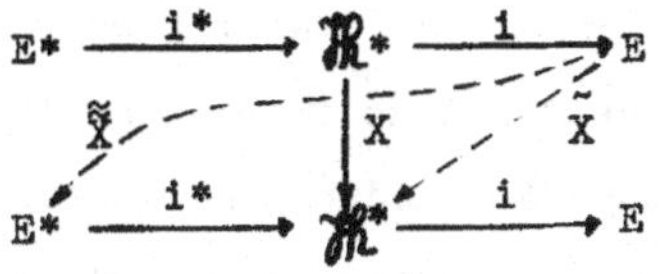

One more example

Let $\mathcal{H}$ be an infinite dimensional complex Hilbert space and let $GL(\mathcal{H}; U(L), Pos(C_2))$ be the subgroup of $GL(\mathcal{H})$ consisting of those operators X such that $[X] = (XX^*)^{\frac{1}{2}}$ lies in $Pos(\mathcal{H}; C_2)$; this Banach-Lie group, of Lie algebra $\underline{u}(\mathcal{H}; L) \oplus Sym(\mathcal{H}; C_2)$, happens to play an important role in the theory of boson fields (Shale [159]).

About the need for ad hoc groups in quantum physics, see also Dieudonné-Bleuler [43].

II.7. - Riemannian geometry on Hilbert-Lie groups

As the previous sections have shown, the groups of operators introduced so far share with the finite dimensional classical Lie groups various properties of an algebraic-topological or differential-topological character. In the cases where these groups are moreover modelled (as manifold) on a Hilbert space, one expect the similarity to carry over to Riemannian-geometrical properties; this is shown to be (at least partially)so in this section. Groups corresponding to L*-algebras play clearly a distinguished role here, which they had no reason to do before.

<u>Definition 7</u>. A <u>Riemann-Hilbert-Lie group</u>, of for short a <u>RHL-group</u>, is a real Banach-Lie group modelled on a Hilbert space and whose Lie algebra is given with a distinguished scalar product. A connected RHL-group will always be considered as furnished with the left invariant Riemannian structure defined by the scalar product on its Lie algebra, and with the corresponding Levi-Cività connection.

Let G be a connected RHL-group with Lie algebra $\underline{g}$, and let $\langle\!\langle \,|\, \rangle\!\rangle$ denote the scalar product on $\underline{g}$. Let $X, Z \in \underline{g}$; the relation

$$\langle\!\langle [X,Y] | Z \rangle\!\rangle = \langle\!\langle B(Z,X) | Y \rangle\!\rangle \text{ for all } Y \in \underline{g}$$

defines a bilinear continuous map $B : \underline{g} \times \underline{g} \longrightarrow \underline{g}$.

Z Geodesics through the identity of G are not necessarily one parameter subgroups, except when B is skew-symmetric!

<u>Proposition 17</u>. Let G be a connected RHL-group and let B be as above.

i) Let ξ and η be two left-invariant vector fields on G; the covariant derivative of η along ξ is given by

$$2\nabla_{\xi}\eta = [\xi,\eta] - B(\xi,\eta) - B(\eta,\xi)$$

ii) The sectional curvature $R_{\xi\eta}$ of G attached to two vector fields ξ and η depends on B only; when B is skew-symmetric:

$$R_{\xi\eta} = \langle\langle [\xi,\eta] \mid B(\xi,\eta) - \tfrac{3}{4}[\xi,\eta] \rangle\rangle$$

<u>Proof</u> : see Arnold [4], [5], or, in the standard case where $B(X,Y) = [X,Y]$, Milnor [120] part IV. ∎

<u>Examples</u>.

i) Any compact Lie group can be given a bi-invariant Riemannian metric, hence can be a fortiori considered as a RHL-group.

ii). More generally, let G be a Banach-Lie group whose Lie algebra is a real L^*-algebra $\underline{g}$. Then G is clearly a RHL-group and the map B is given by $B(X,Y) = [X,-Y^*]$ for all $X,Y \in \underline{g}$. In particular, if $\underline{g}$ is compact (in the sense of definitions I.8 and II.6; $\underline{g}$ might be infinite dimensional), $B(X,Y) = [X,Y]$ for all $X,Y \in \underline{g}$ and the (always positive) sectional curvature is given by the familiar formula $R_{\xi\eta} = \frac{1}{4}\| [\xi,\eta] \|^2$. According to the classification of the separable L^*-algebras, the separable infinite dimensional examples are essentially:

$$U(\mathcal{H};\ C_2) \qquad O^+(\mathcal{H}_{\mathbb{R}};\ C_2) \qquad Sp(\mathcal{H}_{\mathbb{Q}};\ C_2)$$

iii) Let $\underline{g}$ be the Lie algebra of all Hilbert-Schmidt operators on a separable infinite dimensional complex Hilbert space $\mathcal{H}$. The standard norm on $\underline{g}$ is given by the scalar product

$$\langle\langle\ |\ \rangle\rangle \begin{cases} \underline{g} \times \underline{g} \longrightarrow \mathbb{C} \\ (X,Y) \longmapsto \operatorname{trace}(XY^*) \end{cases} ;$$

furnished with it, $\underline{g}$ was denoted by $\underline{gl}(\mathcal{H};\ C_2)$ from section II.5 onwards; the choice of $\langle\langle\ |\ \rangle\rangle$ as scalar product on $\underline{g}$ is natural, as proposition I.8 shows.

However, the Lie algebra $\underline{g}$ can be given other scalar products. One example is as follows : let T be a positive norm-increasing bounded operator on $\mathcal{H}$ and define

$\langle\!\langle \mid \rangle\!\rangle_T : \begin{cases} \underline{g} \times \underline{g} \longrightarrow \mathbb{C} \\ (X,Y) \longmapsto \text{trace}(TXY^*) \end{cases}$. In this way, each operator such as T defines a RHL-structure on the group $GL(\mathcal{H}; C_2)$. An intrinsic study of such structures can be carried along, similar to those of Smiley [165] and Saworotnow [149], [150], who consider non-standard structures on the H*-algebra $C_2(\mathcal{H})$.

II.8.- Remarks, projects and questions

8.1.- There are three separable infinite dimensional compact simple real L*-algebras : $\underline{u}(\mathcal{H}; C_2)$, $\underline{o}(\mathcal{H}_{\mathbb{R}}; C_2)$ and $\underline{sp}(\mathcal{H}_{\mathbb{Q}}; C_2)$. From general principles about universal coverings (Lazard [107], 16.5 and 20.7), each of them corresponds to a unique (up to isomorphism) connected and simply connected Hilbert-Lie group. The group $Sp(\mathcal{H}_{\mathbb{Q}}; C_2)$ is simply connected (section II.6). Though we have not yet written it up in all details, the explicit construction of the universal covering of $O^+(\mathcal{H}_{\mathbb{R}}; C_2)$ could be worked out (see number 8.2 below). But what is the group defined by $\underline{u}(\mathcal{H}; C_2)$, namely the (infinite) universal covering of $U(\mathcal{H}; C_2)$???

8.2.- The universal covering of $O^+(\mathcal{H}_{\mathbb{R}}; C_2)$ will be denoted by $Spin(\mathcal{H}_{\mathbb{R}}; C_2)$. Its construction in finite dimensions is standard (for example : Atiyah-Bott-Shapiro [6] and Karoubi [98]) and can be extended to infinite dimensions by using results from the theory of infinite dimensional Clifford algebras and of the canonical anticommutation relations (Bourbaki [24], Shale-Stinespring [160], [161], Guichardet [69], Slawny [163], [164]). Indeed, the abstract group underlying the Banach-Lie group $Spin(\mathcal{H}_{\mathbb{R}}; C_2)$ is directly related to the group of those canonical transformations which are implementable in all Fock representations of the CAR over the complexification of $\mathcal{H}_{\mathbb{R}}$ (terminology in this last phrase as in [163]). The interested reader should however be warned that our note [78] was relying on a paper [21] which contains an incorrect step in the proof of its lemma 6.

From the abstract definition of Spin($\mathcal{H}_{\mathbb{R}}$; C_2) as a covering group, and using standard techniques of algebraic geometry ([88], 3.1 and 2.10.1), one can easily prove results as in Haefliger [71] and Milnor [121], [122] about the existence of spin-structures on ad hoc Hilbert manifolds.

From the concrete construction of Spin($\mathcal{H}_{\mathbb{R}}$; C_2), it will be possible to define an infinite dimensional analogue of the Dirac operator. Motivations for the interest in infinite dimensional elliptic operators can be found in Dalec'kii [40] and Vishik [197].

8.3.- It would be highly interesting to be able and construct on some Banach-Lie groups measures which would be tied in some sense to the group structure. Consider for example the space $L^2(I \times I)$ of square integrable real valued functions on the product of two unit intervals. This space is isomorphic to $C_2(L^2(I))$, as indicated for example in Ambrose [2] section 1 example 2; hence it has the structure of a L*-algebra to which corresponds a group of the type GL($\mathcal{H}_{\mathbb{R}}$; C_2).

Problem : find a subalgebra $\underline{s}$ of $L^2(I \times I)$ such that the inclusion $\underline{s} \longrightarrow L^2(I \times I)$ be an abstract Wiener space, and translate the Gauss measure so defined on $L^2(I \times I)$ onto the group level. (For abstract Wiener spaces and measures on infinite dimensional manifolds, see Gross [68], Eells-Elworthy [57], Eells [59].)

8.4.- From the explicit knowledge of the root structure of the classical complex Banach-Lie algebra of compact operators, it is elementary to deduce an Iwasawa decomposition for each of the classical real Banach-Lie algebras of compact operators. This can also be done in full generality for separable semi-simple L*-algebras (as in Helgason [84] chap. VI, th. 3.4). A natural further step is to obtain the corresponding global statement. The standard proof ([84], chap. VI section 5) depends on finite dimensional arguments. We conjecture however that such global Iwasawa deompositions exist for the classical real Banach-Lie groups of compact operators; the proof of this conjecture is likely to be through a case-by-case checking.

8.5.- One of the outputs of carrying out the programme sketched in 8.4 would be to furnish several examples of Hilbert-Lie groups which would be nilpotent and solvable. More generally, some theory of nilpotent Banach-Lie algebras is yet to be done; it will hopefully extend the theory or arbitrary finite dimensional real and complex Lie algebras in the same sense that L*-algebras generalise reductive finite dimensional real and complex Lie algebras. We feel that the first thing to do in this direction is a detailed study of groups of quasi-nilpotent operators in Hilbert space. The following example is a very first step.

Example. Let $\mathcal{H}$ be a separable Hilbert space over $\mathbb{R}$ or $\mathbb{C}$. Let $e = (e_n)_{n \in \mathbb{N}}$ be a fixed orthonormal basis of $\mathcal{H}$. Let $\underline{t} = \underline{t}(\mathcal{H}, e; C_2)$ be the Lie algebra of Hilbert-Schmidt operators on $\mathcal{H}$ whose matrix representation with respect to e is strictly upper triangular. The intersection of all the terms in the lower central series of $\underline{t}$ is reduced to zero, hence the "nilpotency" of $\underline{t}$. Let $T = T(\mathcal{H}, e; C_2)$ be the group of operators on $\mathcal{H}$ of the form $id_{\mathcal{H}} + X$, with X strictly upper triangular with respect to e, and X Hilbert-Schmidt. Clearly, T is a sub Banach-Lie group of $GL(\mathcal{H}; C_2)$, and the exponential map $\underline{t} \longrightarrow T$ is given by the traditional power series.

Lemma : Let X be a Hilbert-Schmidt quasi-nilpotent operator on $\mathcal{H}$; then the estimate $\| X^n \|_2 \leq \left((n-1)!\right)^{-\frac{1}{2}} \| X \|_2^{\ n}$ holds for all $n \in \mathbb{N}^*$.

The proof of this lemma, non trivial, is due to Ringrose ([142], th. 5). It follows easily that the series defining $\log(\mathrm{id}_{\mathcal{H}} + X)$ converges in the $C_2(\mathcal{H})$-norm for all $X \in \underline{t}$. Hence the

Proposition : The exponential map $\underline{t} \to T$ is a diffeomorphism. (Recall : if G is a connected, simply connected, finite dimensional nilpotent Lie group, then the exponential map is always a diffeomorphism from the Lie algebra of G onto G.)

Corollary : the exponential map induces on T a structure of Fredholm manifold (see Elworthy-Tromba [62], prop. 1.2).

A starting point for the study of more general "nilpotent" Lie algebras of operators on $\mathcal{H}$ might be found in Schue [155]. See as well Vasilescu [180], Limic [112].

8.6.- Let $\mathcal{A}$ be an associative involutive complex Banach algebra. The answers to the following questions seem to depend strongly on each other : Is the norm topology of $\mathcal{A}$ unique? Are the derivations and the (*-)isomorphisms of $\mathcal{A}$ all continuous? spatial? inner? See Loy [113] and Kadison [96] for an introduction to these problems. Similar questions can be asked for semi-simple (in some sense) Banach-Lie algebras; propositions II.2, II.3, II.9 and II.12 suggest a large proportion of positive answers; idem for Banach-Jordan algebras (see for example [17]).

We recall for memory the projects and conjectures stated pages I.24, II.25 and II.31.

Projects 8.1 to 8.3 are, in my view, the most interesting by far.

Appendix : about semi-simplicity of infinite dimensional Lie algebras

The temporary definition adopted page I.10 was : a Lie algebra is semi-simple if it has no non-trivial abelian ideal. More generally, if $\underline{g}$ is any real or complex Lie algebra, one can define a first notion of a radical as follows (and as in Vasilescu [180]): An ideal $\underline{a}$ in $\underline{g}$ is said to be primitive if any ideal $\underline{b}$ of $\underline{g}$ such that $[\underline{b},\underline{b}] \subset \underline{a}$ is moreover contained in $\underline{a}$, namely such that $\underline{b} \subset \underline{a}$; the radical $R_{\underline{g}}$ of $\underline{g}$ is then the intersection of all primitive ideals of $\underline{g}$. Our temporary definition can then be rephrased as follows: a Lie algebra is semi-simple if its radical $R_{\underline{g}}$ is reduced to zero.

However, this notion is in general not satisfactory. Indeed, there exists an infinite dimensional locally finite locally nilpotent real Lie algebra which is semi-simple according to the definition above:

Example (I. Stewart). Levich [110] gives a locally nilpotent torsion-free group with no subnormal abelian subgroup. By Mal'cev correspondence (e.g. Stewart [170]), it implies that there exists a locally nilpotent Lie algebra over the rationals without abelian subideals; hence ditto over $\mathbb{R}$ or $\mathbb{C}$. The existence of even "worse" examples follows from Levich-Tokarenko [111] and Simonjan [162].

Questions : Is it possible to single out a class of Lie algebras for which the definition above is satisfactory? In particular, is the class of c-involutive Lie algebras (definition I.2) convenient for this purpose?

There are naturally many other ways to define the radical of a Lie algebra, hence to define semi-simplicity: Baer radical, Hirsch-Plotkin radical, Fitting radical; see Stewart [169] part two.

Simple Lie algebras, which were the main concern of Chapter I, are semi-simple with respect to any reasonable definition of semi-simplicity.

We have (temporarily) defined the semi-simplicity of a normed Lie algebra in the same way as in Chapter I (page II.3).

Questions : Is that definition satisfactory for (normed or) Banach-Lie algebras in general ? for c-involutive Banach-Lie algebras ? (it is for L*-algebras).

This question is clearly tied to the project 8.5 (page II.37).

Even if the answer to the later question was affirmative, it would be convenient to have a definition of semi-simplicity in terms of representations. It is indeed possible in finite dimensions (Bourbaki [25] §6, th. 2 and remark 1, and ex. 20) and standard for associative Banach algebras (Rickart [141]).

Appendix : review of norm ideals (Schatten)

Let $\mathcal{H}$ be a separable infinite dimensional complex Hilbert space. If X is an operator on $\mathcal{H}$, [X] denotes its absolute value, that is [X] is the positive operator on $\mathcal{H}$ such $[X]^2 = XX^*$.

Let $\mathcal{I}$ be a non-zero two-sided (associative) ideal in $L(\mathcal{H})$ and let $\alpha : \mathcal{I} \longrightarrow \mathbb{R}$ be a norm. Then α is said to be a crossnorm if $\alpha(X) = \| X \|$ for all operators X of rank one; it is unitarily invariant if $\alpha(VXW) = \alpha(X)$ for all $X \in \mathcal{I}$ and for all $V,W \in U(\mathcal{H})$; it is uniform if $\alpha(YXZ) \leq \| Y \| \alpha(X) \| Z \|$ for all $X \in \mathcal{I}$ and for all $Y,Z \in L(\mathcal{H})$. A norm ideal in $L(\mathcal{H})$ is a non-zero two-sided ideal in $L(\mathcal{H})$ together with a uniform crossnorm with respect to which the ideal becomes a Banach space. A minimal norm ideal in $L(\mathcal{H})$ is a norm ideal in $L(\mathcal{H})$ such that none of its non-trivial closed subspace is again a norm ideal.

A crossnorm on $C_o(\mathcal{H})$ is unitarily invariant if and only if it is uniform. If α is a uniform crossnorm, then $\operatorname{trace}([X]) \geq \alpha(X) \geq \| X \|$ for all $X \in C_o(\mathcal{H})$. If α is a uniform crossnorm on $C_o(\mathcal{H})$, then the map

$$\alpha' : \begin{cases} C_o(\mathcal{H}) \longrightarrow \mathbb{R} \\ X \longrightarrow \sup\left(\dfrac{|\operatorname{trace}(YX)|}{\alpha(Y)}\right) \end{cases}$$

is again a uniform crossnorm said to be <u>associated with α</u> (sup is taken over all non-zero Y's in $C_o(\mathcal{H})$).

<u>Proposition 18</u>. Let α be a uniform crossnorm on $C_o(\mathcal{H})$. Let $L_\alpha(\mathcal{H}) = \{X \in L(\mathcal{H}) \mid \sup\left(\dfrac{|\operatorname{trace}(YX)|}{\alpha'(Y)}\right) < \infty \}$ and let $C_\alpha(\mathcal{H}) = L_\alpha(\mathcal{H}) \cap C(\mathcal{H})$. Then :

i) $C_o(\mathcal{H}) \subset C_\alpha(\mathcal{H}) \subset L_\alpha(\mathcal{H})$ and $C_\alpha(\mathcal{H})$ and $L_\alpha(\mathcal{H})$ are ideals in $L(\mathcal{H})$. If $X \in C_o(\mathcal{H})$, then $\alpha(X) = \sup\left(\dfrac{|\operatorname{trace}(YX)|}{\alpha'(Y)}\right)$.

ii) The evident norm on $L_\alpha(\mathcal{H})$, which is still denoted by α, makes it a norm ideal in $L(\mathcal{H})$ and an involutive Banach algebra.

iii) The restriction of α to $C_\alpha(\mathcal{H})$, which is still denoted by α, makes it a minimal norm ideal and an involutive Banach algebra; conversely, any minimal norm ideal in $L(\mathcal{H})$ is of this type.

iv) Any norm on $L_\alpha(\mathcal{H})$ [resp. on $C_\alpha(\mathcal{H})$] which makes it a Banach algebra is equivalent to α. $L_\alpha(\mathcal{H})$ [resp. $C_\alpha(\mathcal{H})$] is the maximal [resp. minimal] norm ideal in $L(\mathcal{H})$ with respect to α.

v) Let $X \in C_\alpha(\mathcal{H})$ and $Y \in L_{\alpha'}(\mathcal{H})$; then XY and YX are trace-class and trace(XY) = trace (YX). The map

$$\begin{cases} \left(C_o(\mathcal{H}) \text{ furnished with } \alpha\right) \times L_{\alpha'}(\mathcal{H}) \longrightarrow \mathbb{C} \\ (X,Y) \longmapsto \operatorname{trace}(XY) \end{cases}$$

defines an isometric isomorphism of Banach spaces between $L_{\alpha'}(\mathcal{H})$ and the strong dual of the normed space $\left(C_o(\mathcal{H}) \text{ furnished with } \alpha\right)$. Similarly, the strong dual of $C_\alpha(\mathcal{H})$ can be identified with $L_{\alpha'}(\mathcal{H})$.

vi) The following are equivalent :

- $C_\alpha(\mathcal{H})$ is reflexive
- $C_{\alpha'}(\mathcal{H})$ is reflexive
- $L_\alpha(\mathcal{H})$ and $L_{\alpha'}(\mathcal{H})$ are both minimal.

vii) The Hilbert-Schmidt crossnorm is the only uniform crossnorm which is the same as its associate.

Proofs : see among many other possible places Schatten [151], or Dunford-Schwartz ([50] section XI.9), or Gohberg-Krein ([65] chap. III).∎

Example. Let $p \in \bar{\mathbb{R}}$, $1 \leqslant p \leqslant \infty$; the function defined by

$$\| \ \|_p \begin{cases} C_o(\mathcal{H}) \longrightarrow \mathbb{R} \\ X \longmapsto (\mathrm{trace}[X]^p)^{1/p} \end{cases}$$ when $p < \infty$ and by

$\| \ \|_\infty = \| \ \|$ when $p = \infty$ is a uniform crossnorm on $C_o(\mathcal{H})$. The minimal norm ideal defined by $\| \ \|_p$ will be denoted by $C_p(\mathcal{H})$. The associated crossnorm is given by $\| \ \|_q$ where $\frac{1}{p} + \frac{1}{q} = 1$.

If $1 < p < \infty$, $C_p(\mathcal{H})$ is reflexive;

if $p = 1$, $C_1(\mathcal{H}) = L_{\| \ \|_1}(\mathcal{H})$ is the ideal of <u>nuclear</u> operators;

if $p = 2$, $q = 2$ and $C_2(\mathcal{H}) = L_{\| \ \|_2}(\mathcal{H})$ is the ideal of <u>Hilbert-Schmidt</u> operators;

if $p = \infty$, $C_\infty(\mathcal{H}) = C(\mathcal{H})$ is the ideal of <u>compact</u> operators and $L_{\| \ \|_\infty}(\mathcal{H})$ is the trivial ideal of all operators in $L(\mathcal{H})$.

Proposition 19.

i) Let $p, p' \in \bar{\mathbb{R}}$ with $1 \leqslant p \leqslant p' \leqslant \infty$; then $C_p(\mathcal{H}) \subset C_{p'}(\mathcal{H})$ and the canonical injection $C_p(\mathcal{H}) \longrightarrow C_{p'}(\mathcal{H})$ is continuous.

ii) Let $p \in \bar{\mathbb{R}}$ with $2 \leqslant p \leqslant \infty$, let $(e_n)_{n \in N}$ be an orthonormal basis of $\mathcal{H}$ and let X be a bounded operator on $\mathcal{H}$; then $X \in C_p(\mathcal{H})$ if and only if $\sum |Xe_n|^p < \infty$ (the inequality implies that $X \in C_p(\mathcal{H})$ for $1 \leqslant p < 2$).

iii) On $C_1(\mathcal{H})$, the function which associates to an operator X the sum of its eigenvalues (repeated according to multiplicity) is the only continuous linear functional which coincides with the trace on $C_o(\mathcal{H})$.
iv) Let $p \in \bar{\mathbb{R}}$ with $1 < p$; then the space $\underline{sl}(\mathcal{H}; C_o)$ of finite rank operators with zero trace is dense in $C_p(\mathcal{H})$.

Proof. The proofs of i) to iii) are standard [151]. Here is the sketch of a proof for iv).

As any operator in $C_p(\mathcal{H})$ is a linear combination of four positive (hence diagonizable) operators in $C_p(\mathcal{H})$, statement iv) clearly follows from the following lemma. Let $\ell^p_{\mathbb{R}}(N^*)$ be the Banach space of p-summable sequences of real numbers, and let $\mathcal{Z}$ be the subspace consisting of those sequences with finitely many non zero terms only and and whose sum do vanish; then $\mathcal{Z}$ is dense in $\ell^p_{\mathbb{R}}(N^*)$ for all $p \in \bar{\mathbb{R}}$ with $1 < p \leq \infty$. The lemma itself is a straightforward corollary of Hardy's inequality (see for example Rudin [145] chap. 3, exercise 15, with a hint), and can be found in full details elsewhere ([76], lemma 3.3). ∎

Remark. In sections II.5 and II.6, we deal with Banach-Lie algebras corresponding to the particular ideal $C_p(\mathcal{H})$ instead of those corresponding to arbitrary minimal norm ideals $C_\alpha(\mathcal{H})$. This restriction is made only in order to keep the notations reasonably simple. In fact, the case $C_1(\mathcal{H})$ is characteristic of all the cases for which the closure of $\underline{sl}(\mathcal{H}; C_o)$ is one codimensional in $C_\alpha(\mathcal{H})$, and the cases $C_p(\mathcal{H})$ for $1 < p \leq \infty$ are characteristic of all the cases for which $\underline{sl}(\mathcal{H}; C_o)$ is dense in $C_\alpha(\mathcal{H})$.

CHAPTER III.

EXAMPLES OF INFINITE DIMENSIONAL HILBERT SYMMETRIC SPACES

The content of this chapter overlaps substantially with that of a previous note [75]. Hopefully, these examples will be followed sometime by a theory of E. Cartan's symmetric spaces in the context of Banach (or at least Hilbert) manifolds.

Let G be one of the Banach-Lie groups considered in chapter II and let H be a sub Banach-Lie group-manifold of G . Our examples are particular homogeneous spaces of the type $G/_H$. In order to avoid ambiguity, we first recall a standard result (see for example Lazard [107] 23.10).

Result. Let G be a Banach-Lie group with unit e and let H be a sub Banach-Lie group-manifold of G . Then there exists on the left cosets space $G/_H$ a manifold structure, uniquely determined by the condition that the canonical projection $G \xrightarrow{\pi} G/_H$ is a submersion. The canonical action $G \times G/_H \longrightarrow G/_H$ is then smooth.

In particular : if $\underline{g}$ is the Lie algebra of G, if $\underline{h}$ is that of H , and if $\underline{m}$ is a complementary subspace of $\underline{h}$ in $\underline{g}$, then the restriction of $D\pi(e)$ to $\underline{m}$ is an isomorphism between $\underline{m}$ and the tangent space to $G/_H$ at $\pi(e)$.

III.1. - A list of examples

The standard reference for finite dimensional symmetric spaces is Helgason [84].

Among finite dimensional Riemannian manifolds, the class of symmetric spaces plays a considerable role, both for its own sake and as a rich source of examples for more general situations. The theory of the (finite dimensional) Riemannian globally symmetric spaces finds one of its achievements in the famous E.Cartan's classification : any such space (say simply connected) is a product of finitely many irreducible terms, and each term (disregarding Euclidean type) is isometric to a space of a well-known list described as follows.

i) The simply connected compact Lie groups

$SU(n)$, $Spin(n)$ and $Sp(n)$

where n is a positive integer.

ii) The Grassmann manifolds

$$G_k(\mathbb{C}^n) = SU(n)/SU(n) \cap [U(k) \times U(n-k)] \quad \text{(hermitian)}$$

$$SG_k(\mathbb{R}^n) = SO(n)/SO(k) \times SO(n-k) \quad \text{(hermitian if } k = 2)$$

$$G_k(\mathbb{Q}^n) = Sp(n)/Sp(k) \times Sp(n-k)$$

where $k,n \in \mathbb{N}$ with $1 \leqslant k \leqslant n/2$ and where $SG_k(\mathbb{R}^n)$ is the simply connected manifold of oriented k-planes in $\mathbb{R}^n$, which double-covers the manifold $G_k(\mathbb{R}^n)$ of k-planes in $\mathbb{R}^n$.

iii) The manifolds

$$\mathbb{R}\mathbb{C}^+(n) = SU(n)/SO(n)$$

$$\mathbb{C}\mathbb{Q}(n) = Sp(n)/U(n)$$

$$\mathbb{C}\mathbb{R}^+(n) = SO(2n)/U(n)$$

$$\mathbb{Q}\mathbb{C}^+(n) = SU(2n)/Sp(n)$$

where $n \in \mathbb{N}^*$ and where our notations are justified by the following example : Let $CR(n) = {}^{O(2n)}/_{U(n)}$ be the space of orthogonal complex structures on the Euclidean space $\mathbb{R}^{2n}$; then $CR(n)$ has two connected components, one of which being $CR^+(n)$.

iv) The list given in i), ii) and iii) above needs two modifications : first to delete some redundant terms in the case n is an integer smaller than 5 ; second to add a small number of spaces whose description involves the exceptional simple Lie groups. With these modifications, the list contains exactly all (up to isometry) irreducible simply connected Riemannian symmetric spaces of the compact type.

v) The duals of the spaces already described, namely the irreducible Riemannian symmetric spaces of the non-compact type.

Let now $\mathcal{H}_{\mathbb{K}}$ be a $\mathbb{K}$-Hilbert space, separable and infinite dimensional. The classical Banach-Lie groups of compact operators on $\mathcal{H}_{\mathbb{K}}$ (section II.6) make it possible to consider, by analogy, a new list of infinite dimensional homogeneous Banach manifolds. In order to unify the notations, we will write now $Hilb(\mathcal{H}_{\mathbb{K}}; C_p)$ the Banach-Lie group written previously $U(\mathcal{H}_{\mathbb{C}}; C_p)$ [resp. $O(\mathcal{H}_{\mathbb{R}}; C_p)$, $Sp(\mathcal{H}_{\mathbb{Q}}; C_p)$] when $\mathbb{K}$ is $\mathbb{C}$ [resp. $\mathbb{R}$, $\mathbb{Q}$].

Grassmann manifolds as Hilbert manifolds, and their real cohomology rings

Let $k \in \mathbb{N}^*$ and let $\mathbb{K}^k$ be identified to some fixed subspace of $\mathcal{H}_{\mathbb{K}}$. The Grassmann manifold of k-planes in $\mathcal{H}_{\mathbb{K}}$ is given by

$$G_k(\mathcal{H}_{\mathbb{K}}) = \mathrm{Hilb}(\mathcal{H}_{\mathbb{K}}; C_2) \Big/ \mathrm{Hilb}(k) \times \mathrm{Hilb}(\mathcal{H}_{\mathbb{K}} \ominus \mathbb{K}^k ; C_2) ;$$

it is clearly a Hilbert manifold (see section II.7). In the real case, the Grassmann manifold of oriented k-planes is

$$SG_k(\mathcal{H}_{\mathbb{K}}) = O^+(\mathcal{H}_{\mathbb{R}}; C_2) \Big/ SO(k) \times O^+(\mathcal{H}_{\mathbb{R}} \ominus \mathbb{R}^k; C_2) .$$

On the other hand, let $\mathcal{H}_{\mathbb{K}} = \mathcal{H}^-_{\mathbb{K}} \oplus \mathcal{H}^+_{\mathbb{K}}$ be the Hilbert sum of two infinite dimensional Hilbert spaces. Define then

$$G_\infty(\mathcal{H}_{\mathbb{K}}; C_2) = \mathrm{Hilb}(\mathcal{H}_{\mathbb{K}}; C_2) \Big/ \mathrm{Hilb}(\mathcal{H}^-_{\mathbb{K}}; C_2) \times \mathrm{Hilb}(\mathcal{H}^+_{\mathbb{K}}; C_2)$$

and similarly for $SG_\infty(\mathcal{H}_{\mathbb{R}}; C_2)$. Remark that (say when $\mathbb{K} = \mathbb{C}$) the underlying set of $G_\infty(\mathcal{H}; C_2)$ is <u>not</u> that one of the standard Grassmannian $G_\infty(\mathcal{H}; L) = U(\mathcal{H}) / U(\mathcal{H}^-) \times U(\mathcal{H}^+)$. Indeed $G_\infty(\mathcal{H}; L)$ describes all closed subspaces of $\mathcal{H}$ which have both infinite dimension and codimension, while $G_\infty(\mathcal{H}; C_2)$ describes only those which are the image of $\mathcal{H}^-$ by an element of $U(\mathcal{H}; C_2)$.

Similarly, from the groups $\mathrm{Hilb}(\mathcal{H}_{\mathbb{K}}; C_p)$, one defines the manifolds $G_\infty(\mathcal{H}_{\mathbb{K}}; C_p)$ and $SG_\infty(\mathcal{H}_{\mathbb{R}}; C_p)$ for any $p \in \bar{\mathbb{R}}$ with $1 \leqslant p \leqslant \infty$.

It is well-known (see for example McAlpin [114], section IID) that $G_k(\mathcal{H}_{\mathbb{K}})$ is a classifying space for Hilb(k)-principal bundles, and that $G_\infty(\mathcal{H}_{\mathbb{K}}; C_p)$ is a classifying space for $\mathrm{Hilb}(\mathcal{H}_{\mathbb{K}}; C_p)$-principal bundles, hence for $\mathbb{K}$-vector bundles of unbounded finite dimension. This shows the relevance of the spaces just defined and of the corresponding Stiefel manifolds in algebraic topology. However, we will restrict ourselves to recall only the following results.

Proposition 1.

i) The real cohomology of $G_k(\mathcal{H}_{\mathbb{K}})$ is a polynomial algebra over generators of the following even degrees :

$2,4,6,\ldots\ldots,2k$ if $\mathbb{K} = \mathbb{C}$

$4,8,12,\ldots\ldots,4m$ if $\mathbb{K} = \mathbb{R}$ and $k = 2m+1$

$4,8,12,\ldots\ldots,4k$ if $\mathbb{K} = \mathbb{Q}$

$4,8,\ldots,4m-4,2m$ if $\mathbb{K} = \mathbb{R}$ and $k = 2m$.

ii) The real cohomology of $G_\infty(\mathcal{H}_{\mathbb{K}}; C_p)$ is for any p a polynomial algebra over generators of the following degrees :

$2,4,\ldots\ldots,2k,\ldots$ if $\mathbb{K} = \mathbb{C}$

$4,8,\ldots\ldots,4k,\ldots$ if $\mathbb{K} = \mathbb{R}$ or $\mathbb{K} = \mathbb{Q}$.

Proof : these results are classical; see for example Borel [22], section 18 and chap. IV. ∎

Four other Hilbert manifolds and their homotopy types

Let $\mathcal{H}$ be a complex Hilbert space, separable and infinite dimensional, and let $J_{\mathbb{R}}$ be a fixed conjugation on $\mathcal{H}$. The set of all conjugations on $\mathcal{H}$ is then clearly described by the homogeneous space $\mathbb{R}C(\mathcal{H}\, ; L) = U(\mathcal{H})/O(\mathcal{H}_{\mathbb{R}})$, where $\mathcal{H}_{\mathbb{R}}$ is the real Hilbert space defined by the fixed points of $J_{\mathbb{R}}$ in $\mathcal{H}$. Define a new Hilbert manifold by

$$\mathbb{R}C^+(\mathcal{H}; C_2) = U(\mathcal{H}\, ; C_2) \Big/ O^+(\mathcal{H}_{\mathbb{R}}; C_2) \quad ;$$

it can be referred to as the connected component of the space of those conjugations on $\mathcal{H}$ which are "Hilbert-Schmidt perturbations" of the given $J_{\mathbb{R}}$. In the same way, one defines

$$\mathbb{C}\mathbb{Q}(\mathcal{H}\,;\,C_2) = Sp(\mathcal{H}_{\mathbb{Q}};\,C_2)\Big/U(\mathcal{H}\,;\,C_2)$$

$$\mathbb{C}\mathbb{R}^+(\mathcal{H}\,;\,C_2) = O^+(\mathcal{H}_{\mathbb{R}};\,C_2)\Big/U(\mathcal{H}\,;\,C_2)$$

$$\mathbb{Q}\mathbb{C}(\mathcal{H}\,;\,C_2) = U(\mathcal{H}\,;\,C_2)\Big/Sp(\mathcal{H}_{\mathbb{Q}};\,C_2) \quad .$$

Similarly for the Banach manifolds $\mathbb{R}\mathbb{C}^+(\mathcal{H}\,;\,C_p)$,..., for any $p \in \overline{\mathbb{R}}$ with $1 \leqslant p \leqslant \infty$.

The homotopy type of these spaces is again well-known. We will treat in detail the first example only; the same method applies to the other cases above, and indeed to many more.

Let $\mathcal{H}$, $J_{\mathbb{R}}$ and $\mathcal{H}_{\mathbb{R}}$ be as above, and let $e = (e_n)_{n \in N^*}$ be an orthonormal basis of $\mathcal{H}_{\mathbb{R}}$; then e is a fortiori an orthonormal basis of $\mathcal{H}$. For each $n \in N^*$, identify $\mathbb{R}^n$ [resp. $\mathbb{C}^n$] to the subspace of $\mathcal{H}_{\mathbb{R}}$ [resp. $\mathcal{H}$] span by the first n basis vectors. The group $SO(n)$ [resp. $U(n)$] is accordingly identified to a subgroup of $O^+(\mathcal{H}_{\mathbb{R}};\,C_p)$ [resp. $U(\mathcal{H};\,C_p)$]. Let $SO(\infty)$ denote the limit of the closed expanding system (see Hansen [72]) defined by the groups $SO(n)$'s, and similarly for $U(\infty) = \varinjlim U(n)$ and $\mathbb{R}\mathbb{C}^+(\infty) = U(\infty)/SO(\infty) = \varinjlim U(n)/SO(n)$. Consider the diagram

$$\begin{array}{ccccccc}
SO(n) & \longrightarrow & SO(n+1) & \dashrightarrow & SO(\infty) & \xrightarrow{\alpha} & O^+(\mathcal{H}_{\mathbb{R}};\,C_p) \\
\downarrow & & \downarrow & & \downarrow & & \downarrow \\
U(n) & \longrightarrow & U(n+1) & \dashrightarrow & U(\infty) & \xrightarrow{\beta} & U(\mathcal{H}\,;\,C_p) \\
\downarrow & & \downarrow & & \downarrow & & \downarrow \\
\mathbb{R}\mathbb{C}^+(n) & \longrightarrow & \mathbb{R}\mathbb{C}^+(n+1) & \dashrightarrow & \mathbb{R}\mathbb{C}^+(\infty) & \xrightarrow{\gamma} & \mathbb{R}\mathbb{C}^+(\mathcal{H}\,;\,C_p)
\end{array}$$

All the "vertical" triples are Serre fibrations, so that there are homotopy long exact sequences associated to each of them. In particular, for each $i \in N^*$, the two rows of the diagram

$$\begin{array}{ccccc}
\pi_i(SO(\infty)) & \longrightarrow & \pi_i(U(\infty)) & \longrightarrow & \pi_i(\mathbb{R}\mathbb{C}^+(\infty)) \longrightarrow \\
\downarrow \alpha_i & & \downarrow \beta_i & & \downarrow \gamma_i \\
\pi_i(O^+(\mathcal{H}_{\mathbb{R}};\, C_p)) & \longrightarrow & \pi_i(U(\mathcal{H};\, C_p)) & \longrightarrow & \pi_i(\mathbb{R}\mathbb{C}^+(\mathcal{H};\, C_p)) \longrightarrow
\end{array}$$

$$\begin{array}{ccc}
\longrightarrow \pi_{i-1}(SO(\infty)) & \longrightarrow & \pi_{i-1}(U(\infty)) \\
\downarrow \alpha_{i-1} & & \downarrow \beta_{i-1} \\
\rightarrow \pi_{i-1}(O^+(\mathcal{H}_{\mathbb{R}};\, C_p)) & \longrightarrow & \pi_{i-1}(U(\mathcal{H};\, C_p))
\end{array}$$

are exact. As α and β are homotopy equivalences (corollary to proposition II.16), α_i and β_i are isomorphisms. It follows from the five lemma (see e.g. Cartan-Eilenberg [33] chap. I, §1) that γ_i is a group isomorphism for all $i \in N$ (the case $i = 0$ is trivial) ; in other words, γ is a weak homotopy equivalence. But $\mathbb{R}\mathbb{C}^+(\infty)$ is homotopically equivalent to an ANR (Hansen [72] cor. 6.4) and the Banach manifold $\mathbb{R}\mathbb{C}^+(\mathcal{H};\, C_p)$ is an ANR. Hence Whitehead's lemma applies (see e.g. Palais [130] section 6.6) and γ is a homotopy equivalence. We have proved :

<u>Proposition 2</u>. Notations being as above, the four following inclusion maps are homotopy equivalences for all $p \in \bar{\mathbb{R}}$ with $1 \leq p \leq \infty$:

$$\mathbb{R}\mathbb{C}^+(\infty) \longrightarrow \mathbb{R}\mathbb{C}^+(\mathcal{H};\, C_p)$$
$$\mathbb{C}\mathbb{Q}(\infty) \longrightarrow \mathbb{C}\mathbb{Q}(\mathcal{H};\, C_p)$$
$$\mathbb{C}\mathbb{R}^+(\infty) \longrightarrow \mathbb{C}\mathbb{R}^+(\mathcal{H};\, C_p)$$
$$\mathbb{Q}\mathbb{C}(\infty) \longrightarrow \mathbb{Q}\mathbb{C}(\mathcal{H};\, C_p)$$

Examples of Hilbert (and Banach) manifolds corresponding to the irreducible Riemannian symmetric spaces of the <u>non-compact type</u> can be obtained the same way.

III.2. - On symmetric spaces

The classical theory of E. Cartan's symmetric spaces takes a constant advantage of the close relationship between the three following notions : orthogonal symmetric Lie algebra, Riemannian symmetric pair, and Riemannian (globally) symmetric space. We check now that these notions still make sense for the manifolds listed in section III.1. The various cases are sufficiently similar to each other for us to deal with one example only. We follow again the terminology of Helgason [84] .

Let $\mathcal{H} = \mathcal{H}^- \oplus \mathcal{H}^+$ be the direct sum of two infinite dimensional separable real Hilbert spaces. Consider the Hilbert Riemannian manifold

$$M = SG_\infty(\mathcal{H} ; C_2) = O^+(\mathcal{H} ; C_2) \Big/ O^+(\mathcal{H}^-; C_2) \times O^+(\mathcal{H}^+; C_2) .$$

Put $G = O^+(\mathcal{H}; C_2)$ and $K = O^+(\mathcal{H}^-; C_2) \times O^+(\mathcal{H}^+; C_2)$; the Lie algebras of these Banach-Lie groups are respectively $\underline{g} = \underline{o}(\mathcal{H}; C_2)$ and $\underline{k} = \underline{o}(\mathcal{H}^-; C_2) \times \underline{o}(\mathcal{H}^+; C_2)$. Let I be the orthogonal operator on $\mathcal{H}$ which is equal to minus the identity on $\mathcal{H}^-$ and to the identity on $\mathcal{H}^+$. Then the map $\sigma\begin{cases} G \longrightarrow G \\ X \longmapsto IXI \end{cases}$ is an involutive analytic automorphism of G ; let $s : \underline{g} \longrightarrow \underline{g}$ be its derivative at the identity e of G . The set of fixed points of s is $\underline{k}$ and its eigenspace for the eigenvalue -1 is $\underline{m} = (C_2(\mathcal{H}^- \to \mathcal{H}^+) \oplus C_2(\mathcal{H}^+ \to \mathcal{H}^-)) \cap \underline{g}$; note that $\underline{m}$ is isomorphic as a Hilbert space to $C_2(\mathcal{H})$.

The object $(\underline{g}, s)$ should obviously be called an <u>orthogonal symmetric Lie algebra</u>, of the compact type and irreducible ([84] , pages 178, 194 and 306). The pair (G,K) should be called a <u>Hilbert Riemannian symmetric pair</u>, of the compact type and irreducible ([84], pages 174, 195 and 306), which is clearly associated with $(\mathfrak{g}, s)$.

Let now π be the canonical projection $G \longrightarrow M$, let e be the identity in G and let ε be the image $\pi(e)$. Define a map $s_\varepsilon : M \longrightarrow M$ by the condition $s_\varepsilon \circ \pi = \pi \circ \sigma$. Let now $x = bK$ be an arbitrary point in M , where bK is the coset corresponding to some b in G ; let λ_b be the diffeomorphism of M defined by b and put $s_x = \lambda_b \circ s_\varepsilon \circ \lambda_b^{-1}$ (s_x depends on the coset bK only, not on b). It is easy to check that s_x is then an involutive isometry of M with x as isolated fixed point. In other words, M is a <u>Hilbert Riemannian globally symmetric space</u>.

Finally, all considerations concerning the <u>duality</u> of M with the manifold

$$O^+(\mathcal{H}, \infty, \infty; C_2)/O^+(\mathcal{H}^-; C_2) \times O^+(\mathcal{H}^+; C_2)$$

can now be repeated precisely as in Helgason [84], chap. V §2, example I.

Many basic geometrical properties of finite dimensional symmetric spaces carry over to the spaces introduced in section III.1 and to cartesian products of them. For example, furnished with their natural Riemannian metric, these Hilbert manifolds have positive (resp. negative) curvature if they are of the compact (resp. non-compact) type. It follows among other things that, for those of the non-compact type, the exponential map at some point is a diffeomorphism (McAlpin [114], section IH); hence, via the Elworthy-Tromba construction, these spaces have distinguished Fredholm structures ([62] , proposition 1.2).

Instead of dealing further with this kind of general statements, we prefer now to consider in some detail a specific example. It was studied, in finite dimensions, by Mostow [124]; his proves carry over to the Hilbert case with minor modifications only.

Let $\mathcal{H}$ be an infinite dimensional separable real Hilbert space. Put $G = GL(\mathcal{H} ; C_2)$, $K = O(\mathcal{H} ; C_2)$, let $S = Sym(\mathcal{H} ; C_2)$ be the subspace of the self-adjoints operators in $C_2(\mathcal{H})$ and let $P = Pos(\mathcal{H} ; C_2)$ be the set of positive definite operators of the form $id_{\mathcal{H}} + X$ with $X \in C_2(\mathcal{H})$. The set P is furnished with a structure of Hilbert manifold by the obvious bijection between P and the open subset of S containing those X whose spectrum lies in the real interval $]-1, \infty[$.

The group G acts smoothly and transitively on the manifold P by $\begin{cases} G \times P \longrightarrow P \\ (X,A) \longmapsto X^*AX \end{cases}$, and the isotropy subgroup at $id_{\mathcal{H}}$ is clearly K. The induced map $\begin{cases} G \longrightarrow P \\ X \longmapsto X^*X \end{cases}$ is a submersion; indeed, its derivative at some point X in G is $Df(X) \begin{cases} C_2(\mathcal{H}) \longrightarrow S \\ Y \longmapsto X^*Y + Y^*X \end{cases}$, and is the composition of the isomorphism from $C_2(\mathcal{H})$ into itself given by $Y \longmapsto X^*Y$ and of the surjective map with splitting kernel from $C_2(\mathcal{H})$ to S given by $Z \longmapsto Z + Z^*$. It follows that the natural map $G/K \longrightarrow P$ is an analytic diffeomorphism of Hilbert manifolds (see the result recalled page III.1).

Let $A \in P$ and let S,T be two tangent vectors at P to A. The tangent space at P to A being identified with S, one can define a smooth Riemannian structure on P by

$$\langle S|T\rangle_A = \mathrm{trace}(A^{-1}SA^{-1}T) \quad ;$$

this structure is clearly G-invariant.

<u>Proposition 3.</u>

i) Two points A and B in P can always be joined by a geodesic of minimal length, which is unique and given as follows : let $X \in P$ be such that $X^*X = A$ and let $Y = X^{*-1}BX^{-1}$; then the geodesic is $\begin{cases} [0,1] \longrightarrow P \\ t \longmapsto X^* \exp(t \log Y) X \end{cases}$.

ii) The curvature is smaller than or equal to zero at all points of P ; the sum of the angles of a geodesic triangle in P is always smaller than or equal to π .

iii) The exponential map $S \longrightarrow P$ is a diffeomorphism, which is conformal at the origin of S .

Definition. Let $\underline{m}$ be a subspace of a Banach-Lie algebra $\underline{g}$. A Lie triple system in $\underline{m}$ is a closed subspace $\underline{t}$ of $\underline{m}$ such that $[\underline{t}, [\underline{t},\underline{t}]] \subset \underline{t}$.

Proposition 4. Let $\mathcal{E}$ be a closed linear subspace of S and let E be the image of $\mathcal{E}$ by the exponential map; then the following are equivalent :

i) E is a geodesic subspace of P

ii) for all $e,f \in E$, $efe \in E$

iii) for all $S,T \in \mathcal{E}$, $[S, [S,T]] \in \mathcal{E}$

iv) $\mathcal{E}$ is a Lie triple system in S .

Proof of propositions 3 and 4 : as in Mostow [124] . ∎

III.3. - Poincaré series

This section describes an apparently eccentric fact. It can be read as a challenge to whoever will find of it a respectable interpretation. It shows an unexpected connection between the cohomology of the spaces of section III.1 and the classical domain of modular functions.

Let M be a topological space (say, up to homotopy, an ANR, in order to avoid complications). For each $k \in N$, let b_k be the kth Betti number of M, and suppose M such that all the b_k's are finite. Then the Poincaré series of M is, by definition, the formal power series $P_M(t) = \sum_{k \in N} b_k t^k$; it clearly depends on the homotopy type of M only. We want now to compute P_M when M is a classical Banach-Lie group of compact operators or when M is one of the spaces introduced in section III.1.

Recall of some formal power series P_M

The spaces we are interested in are all homotopically equivalent to stable spaces, in the sense one speaks of the stable classical groups (results as the corollary to prop. II.16, or as proposition III.2). Their Betti number can easily be found in the literature.

Classical groups. The various ways of computing P_M when M is a finite dimensional classical Lie group are reviewed by Samelson in [148]. What we need can be summed up as follows (see again the corollary to prop. II.16).

The Poincaré series of $U(\infty)$, or of $U(\mathcal{H}; C_p)$, is given by

$$P_U(t) = \prod_{k=1}^{\infty} (1 + t^{2k-1}) .$$

The Poincaré series of $SO(\infty)$, or of $O^+(\mathcal{H}_{\mathbb{R}}; C_p)$, is given by $P_O(t) = \prod_{k=1}^{\infty} (1 + t^{4k-1})$. The Poincaré series of $Sp(\infty)$, or of $Sp(\mathcal{H}_{\mathbb{Q}}; C_p)$, is given by $P_{Sp}(t) = P_O(t)$.

Grassmannian. From the results recalled in proposition III.1, it is clear that P_M is a polynomial whenever M is a Grassmannian $G_k(\mathcal{H}_{\mathbb{K}})$ with k finite, and that one has :

$$\text{for } G_\infty(\mathcal{H}; C_p) \quad : \quad P_{G\mathbb{C}}(t) = \prod_{k=1}^{\infty} (1 - t^{2k})^{-1}$$

$$\text{for } G_\infty(\mathcal{H}_{\mathbb{R}}; C_p) \quad : \quad P_{G\mathbb{R}}(t) = \prod_{k=1}^{\infty} (1 - t^{4k})^{-1}$$

$$\text{for } G_\infty(\mathcal{H}_{\mathbb{Q}}; C_p) \quad : \quad P_{G\mathbb{Q}}(t) = P_{G\mathbb{R}}(t) .$$

Other spaces. The Poincaré polynomials of the finite dimensional manifolds $\mathbb{R}\mathbb{C}^+(n)$, $\mathbb{C}\mathbb{Q}(n)$, $\mathbb{C}\mathbb{R}^+(n)$ and $\mathbb{Q}\mathbb{C}^+(n)$ can be computed with the aid of Hirsch formula (see Borel [22] section 20, particularly theorem 6a). From that, and using proposition III.2, it follows easily that one has :

$$\text{for } \mathbb{R}\mathbb{C}^+(\mathcal{H}; C_p) \quad : \quad P_{\mathbb{R}\mathbb{C}}(t) = \prod_{k=1}^{\infty} (1 + t^{4k+1})$$

$$\text{for } \mathbb{C}\mathbb{Q}(\mathcal{H}; C_p) \quad : \quad P_{\mathbb{C}\mathbb{Q}}(t) = \prod_{k=1}^{\infty} (1 + t^{2k})$$

$$\text{for } \mathbb{C}\mathbb{R}^+(\mathcal{H}; C_p) \quad : \quad P_{\mathbb{C}\mathbb{R}}(t) = P_{\mathbb{C}\mathbb{Q}}(t)$$

$$\text{for } \mathbb{Q}\mathbb{C}(\mathcal{H}; C_p) \quad : \quad P_{\mathbb{Q}\mathbb{C}}(t) = P_{\mathbb{R}\mathbb{C}}(t) .$$

The Poincaré series of the other classical Banach-Lie groups of compact operators are polynomials or are similar to one of $P_U(t)$ and $P_O(t)$, according to the corollary to proposition II.16. The Hilbert manifolds corresponding to the Riemannian symmetric space of the non-compact type are of no interest in this section, because these spaces are contractible (see Helgason [84], chap. VI, theorem 1.1.iii).

Recall of infinite products and of Jacobi functions

The infinite products written above have been first introduced by Euler, in the domain of number theory [63]. The relationship between the Poincaré polynomials of, say, the Grassmannians, and combinatorial properties of integer numbers is standard; it is shown by the method of Schubert's cells for the computation of the cohomology rings of these manifolds (see for example Chern [35], chap. IV, section 1).

On the other hand, let $\theta_1, \ldots, \theta_4$ be the four Jacobi Theta-functions; notations are as in Whittaker-Watson [184], or Bateman [19]. Let λ be the modular function defined by

$$\lambda(t) = \left[\frac{\theta_2(0,t)}{\theta_3(0,t)}\right]^4 = 1 - \left[\frac{\theta_4(0,t)}{\theta_3(0,t)}\right]^4$$

For the properties of λ, see [19] section 13.24; an elementary introduction is in [147], chap. VII, §11.

We need now the formulas

$$\prod_{k=1}^{\infty} (1 + t^{2k-1})^{24} = 16\, t \left[\frac{\theta_3(0,t)}{\theta_2(0,t)} \frac{\theta_3(0,t)}{\theta_4(0,t)}\right]^4$$

and

$$\prod_{k=1}^{\infty} (1 + t^{2k})^{24} = \frac{1}{256t^2} \left[\frac{\theta_2(0,t)}{\theta_3(0,t)}\right]^8 \left[\frac{\theta_3(0,t)}{\theta_4(0,t)}\right]^4$$

as well as

$$\prod_{k=1}^{\infty} (1 - t^{2k})^{12} = \frac{[\theta_1'(0,t)]^4}{16t}$$

(the dash ' in the last formula holds for derivation with respect to the first variable).

Indications for the proofs of these equalities can be found in Whittaker-Watson [184]; see example 10 following section 21.9 for the first two, and section 21.42 for the third.

Poincaré series

It is easy to see that all the infinite products written in this section converge in the unit disc of the complex plane, and that they define in this domain holomorphic functions. By juxtaposition of the facts recalled so far, it is elementary to check that these functions are given as follows (the function defined by the product $P_U(t)$ is still denoted the same way) :

$$[P_U(t)]^{24} = \frac{16t}{\lambda(t)\,[1-\lambda(t)]}$$

$$P_{GC}(t)^{12} = \frac{16t}{[\theta_1'(0,t)]^4}$$

$$P_{GR}(t) = P_{GQ}(t) = P_{GC}(t^2)$$

$$[P_{CQ}(t)]^{24} = \frac{\lambda^2(t)}{256t^2\,[1-\lambda(t)]}$$

$$P_{CR}(t) = P_{CQ}(t)$$

We have not yet been able to express in terms of standard functions either $P_O = P_{Sp}$ or $P_{RC} = P_{QC}$. This reduces clearly to one problem, because of the functional relation given by $(1+t)P_O(t)P_{RC}(t) = P_U(t)$, which is trivially checked from the infinite product expansions.

We will sum up this section by a result and a problem.

Result. Let M be a space which is (up to homotopy equivalence) either a stable classical group, or a stable irreducible globally symmetric space. Then the Poincaré series of M defines a holomorphic complex function (still denoted by P_M) in the open unit disc of the complex plane. These functions are either polynomials, or surprisingly simply related to modular functions.

Remark. The stable spaces defined by the irreducible hermitian symmetric spaces of the compact type are $G_k(\mathbb{C}^\infty) = \varinjlim_n G_k(\mathbb{C}^n)$ for $k \geqslant 1$, $G_2(\mathbb{R}^\infty) = \varinjlim G_2(\mathbb{R}^n)$, $CQ(\infty)$ and $CR^+(\infty)$. For them and for $U(\infty)$, the above result can be expressed more precisely : if P_M is the function defined by the Poincaré series of one of these spaces, then P_M is the product of a rational function by a modular function.

Problem. Express the function defined by $P_0(t) = \prod_{k=1}^{\infty} (1 + t^{4k-1})$ in terms of known functions.

About particular functions defined by Poincaré series of classifying spaces, see Venkov [181] and Quillen [137] .

III.4. - Miscellaneous

Hilbert manifolds of constant curvature (Michal [119], McAlpin [114] and of section IH, Unsain [176])

As in finite dimensions, spaces of constant curvature are symmetric spaces. More precisely, the following holds.

Let M be a simply connected complete Riemannian manifold of constant curvature, modelled on an infinite dimensional separable real Hilbert space $\mathcal{H}$. Then M is isometric to a sphere $\mathcal{H}(r) = \{x \in \mathcal{H} \mid |x|^2 = r^2\}$ if the curvature of M is positive, to $\mathcal{H}$ itself if the curvature of M vanishes, and to a hyperbolic space
$Hyp(r) = \{\xi \oplus x \in R \oplus \mathcal{H} \mid -|\xi|^2 + |x|^2 = -r^2\}$ if the curvature of M is negative. In particular, M is diffeomorphic to one of

$$\mathcal{H}(1) = O^+(\mathcal{H};\, C_2) \big/ O^+(\mathcal{H} \oplus \mathbb{R};\, C_2) \qquad \text{(curvature +1)}$$

$$\mathcal{H} \qquad \text{(curvature 0)}$$

$$Hyp(1) = O^+(\mathcal{H},1,\infty;\, C_2) \big/ O^+(\mathcal{H} \oplus \mathbb{R};\, C_2) \qquad \text{(curvature -1)}$$

furnished with their standard Riemannian structures.

Various examples of Hilbert manifolds are given by the quotients of these three manifolds by the standard discrete subgroups of isometries (see Wolf [185] section 2.7). Each of these quotients is an Eilenberg-McLane space, because their universal coverings written just above are contractible.

An explicit trivialisation of the tangent bundle to the unit sphere in a Hilbert space

The unit sphere $\mathcal{H}(1)$ in the infinite dimensional real Hilbert space $\mathcal{H}$ being contractible, its tangent bundle $T(\mathcal{H}(1))$ is obviously trivial. We indicate below an <u>explicit</u> trivialisation. Let $\mathbb{R}^\infty$ be the inductive limit of the vector spaces R^n's, canonically embedded in each other; consider $\mathbb{R}^\infty$ as a subspace of the Hilbert space $\mathcal{H} = \ell^2$. Zvengrowski ([191], sections 3 and 4) has constructed a norm-preserving multiplication orthogonal to the identity on $\mathbb{R}^\infty$, that is a linear map $\mu: \mathbb{R}^\infty \otimes \mathbb{R}^\infty \longrightarrow \mathbb{R}^\infty$ such that

i) $|\mu(x \otimes y)| = |x||y|$ for all $x, y \in \mathbb{R}^\infty$,

ii) $\mu(e_o \otimes x) = x$ for all $x \in \mathbb{R}^\infty$, where e_o is the first element of the canonical basis of ℓ^2.

iii) $\langle y | \mu(x \otimes y) \rangle = 0$ for all $x, y \in R^\infty$ with $x \perp e_o$.

The unique continuous extention $\bar{\mu}$ of μ to ℓ^2 makes ℓ^2 an absolute valued (not associative) real algebra. For each $x \in \ell^2$, the left-multiplication operator $L_x : y \longmapsto \bar{\mu}(x \otimes y)$ is continuous and invertible, while the right-multiplication operator R_x is not onto in general. Furnished with the multiplication $\bar{\mu}$, the space ℓ^2 could therefore be called a "normal real left-division algebra with a left-unit".

Let now $(e_n)_{n \in N}$ be the canonical basis of ℓ^2, and let μ_j be the vector field on $\mathcal{H}(1)$ defined by $\mu_j(x) = \bar{\mu}(e_j \otimes x)$ for all $x \in \mathcal{H}(1)$, for all $j \in N^*$. Then, for each $x \in \mathcal{H}(1)$, the family $(\mu_j(x))_{j \in N^*}$ is an orthonormal basis of the tangent space to $\mathcal{H}(1)$ at x. This provides an explicit trivialisation of $T(\mathcal{H}(1))$.

The question of the existence of a division algebra structure on ℓ^2 is, to my knowledge, unsolved. The answer has been conjectured to be negative by Wright ([186], page 332). It is known that, if such an algebra existed, it could not be associative (Gel'fand-Mazur theorem), and even not alternative (Bruck and Kleinfeld [29]).

In connection with the study of Fredholm structures on the projective space defined by $\mathcal{H}(1)$, the following question can be asked.

Problem. Does there exist a bilinear map $b : \mathcal{H} \times \mathcal{H} \longrightarrow \mathcal{H}$ having the properties i and ii below?

i) symmetry : $b(x,y) = b(y,x)$ for all $x,y \in \mathcal{H}$.

ii) the left-multiplication by $x : y \longmapsto b(x,y)$ is a Fredholm operator for all $x \in \mathcal{H}$, $x \neq 0$.

CHAPTER IV.

ON THE COHOMOLOGY OF THE CLASSICAL COMPLEX LIE ALGEBRAS OF COMPACT OPERATORS

Let U be a (finite dimensional, simple, compact) classical group, let G be its complexification, let $\underline{u}$ be the Lie algebra of U and let $\underline{g}$ be that of G. The following facts are well-known (Cartan [31], Chevalley-Eilenberg [38], Koszul [100]) :

The real (or de Rham) cohomology $H^*(U)$ of the group U, the cohomology $H^*(\underline{u})$ of its Lie algebra and the algebra $J^*(\underline{u})$ of invariant cocycles on $\underline{u}$ are isomorphic to each other. Moreover, $J^*(\underline{u})$ is an exterior algebra over the space $P^*(\underline{u})$ of primitive cocycles, which is of dimension ℓ (ℓ = rank of U). The same is true for the complex algebras $H^*(G)$, $H^*(\underline{g})$ and $J^*(\underline{g})$; indeed, the inclusion of U in G is a homotopy equivalence, and the relations $H^*(\underline{g}) = H^*(\underline{u}) \otimes_{\mathbb{R}} \mathbb{C}$ and $J^*(\underline{g}) = J^*(\underline{u}) \otimes_{\mathbb{R}} \mathbb{C}$ are true by general principles.

Let now $I(U)$ be the Z-graded algebra of those polynomial functions on $\underline{u}$ with real values which are invariant by the adjoint action of U on $\underline{u}$. Let $H^*(B_U)$ be the real cohomology of the classifying space B_U for U-principal bundles. Then $I(U)$ and $H^*(B_U)$ are isomorphic to each other. Moreover, $I(U)$ is a polynomial algebra over ℓ generators.

Finally, there exists a canonical linear map T (denoted by ρ in Cartan [31]) from $I(U)$ to $J^*(\underline{u})$ which transforms generators of $I(U)$ into generators of $J^*(\underline{u})$. This map plays a crucial role in the study of transgressions.

The purpose of this chapter is to investigate the analogue propositions when $\underline{g}$ is replaced by a classical complex Lie algebra of compact operators. Indications have been sketched in [82], [83]. In the absence of any general theorem, our only method is to perform explicit computations.

IV.1. - The algebra $J_c^*(\underline{g})$ of invariant cochains

This section is an easy consequence of the (classical) computations written up in [81].

Let $\underline{g}$ be a (possibly infinite dimensional) Lie algebra over $\mathbb{K}$ ($\mathbb{K}$ is $\mathbb{R}$ or $\mathbb{C}$). If $k \in \mathbb{N}^*$, a <u>k-cochain</u> is a multilinear alternating map $\underline{g} \times \ldots \times \underline{g}$ (k times) $\longrightarrow \mathbb{K}$. The vector space of all k-cochains on $\underline{g}$ will be denoted by $C^k(\underline{g})$ and the algebra of all cochains on $\underline{g}$ by $C^*(\underline{g}) = \bigoplus_{k \in \mathbb{N}} C^k(\underline{g})$; according to the usual convention, $C^o(\underline{g}) = \mathbb{K}$. The standard explicit formulas for the Lie derivative θ^* and for the "exterior" differentiation d carry over to the infinite dimensional case without change, so that we will not define every notion in detail; see for example Koszul [100]. We will denote by $Z^k(\underline{g})$ [resp. $B^k(\underline{g})$, $H^k(g)$, $J^k(\underline{g})$] the space of the k-cocycles [resp. k-boundaries, k-cohomology classes, k-invariant cocycles] on $\underline{g}$.

Let now $\underline{g}$ be a Banach-Lie algebra over $\mathbb{K}$. The space of <u>continuous k-cochains</u> on $\underline{g}$ will be denoted by $C_c^k(\underline{g})$; similarly for $Z_c^k(\underline{g})$, $B_c^k(\underline{g})$, $H_c^k(\underline{g})$ and $J_c^k(\underline{g})$. We now proceed to describe $J_c^*(\underline{u})$ when $\underline{u}$ is the compact form of a classical complex Lie algebra $\underline{g}$ of compact operators on the <u>infinite dimensional separable complex</u> Hilbert space $\mathcal{H}$.

Consider first the associative algebra $C_o(\mathcal{H})$. For each $k \in \mathbb{N}^*$, define the k-linear alternating map

$$q_k : \begin{cases} C_o(\mathcal{H}) \times \ldots \times C_o(\mathcal{H}) \longrightarrow \mathbb{C} \\ (X_1, \ldots\ldots, X_k) \longmapsto \operatorname{trace}\left(\sum_\sigma sg(\sigma)\, X_{\sigma(1)} \cdot \ldots \cdot X_{\sigma(k)}\right) \end{cases}$$

where the sum is taken over all permutations of the symmetric group in k variables $\mathcal{S}_k$.

It is easy to check that q_k vanishes whenever k is even.

Let then $\underline{u}_o$ denote for short the Lie algebra of finite rank operators $\underline{u}(\mathcal{H}; C_o)$ and define for each $k \in N^*$

$$\hat{\rho}_k \quad \begin{cases} \underline{u}_o \times \dots \times \underline{u}_o \longrightarrow \mathbb{R} \\ (X_1, \dots, X_{2k-1}) \qquad i^{k(2k-1)} q_{2k-1}(X_1, \dots, X_{2k-1}). \end{cases}$$

Proposition 1A. The cochains $(\hat{\rho}_k)_{k \in N^*}$ are primitive cocycles on $\underline{u}_o$. The algebra $J^*(\underline{u}_o)$ is an exterior algebra generated by $(\hat{\rho}_k)_{k \in N^*}$; moreover, these primitive generators are unique up to multiplication by non-zero real numbers.

Proof.

Step one : Let F be an n-dimensional subspace of $\mathcal{H}$ and let $\underline{u}(F)$ be the subalgebra of $\underline{u}_o$ consisting of those operators which map F into itself and its orthogonal complement onto zero. Let $\hat{\rho}_k^F$ be the restriction of $\hat{\rho}_k$ to $\underline{u}(F)$ for all $k \in N^*$. Then $(\hat{\rho}_1^F, \dots, \hat{\rho}_n^F)$ are primitive cocycles on $\underline{u}(F)$, they generate $J^*(\underline{u}(F))$, and they are the unique primitive generators of $J^*(\underline{u}(F))$ up to multiplication by a non-zero real numbers. The proof of this step is classical; the explicit form of the $\hat{\rho}_k^F$'s is apparently due to Dynkin [52]; references and pedestrian computations have been collected in [81].

Step two. Let γ be an invariant ℓ-cochain on $\underline{u}_o$. According to the results recalled under step one, for any subspace F of $\mathcal{H}$ as above, there are constants (a priority depending on F) such that the restriction of γ to $\underline{u}(F)$ can be written as

$$\gamma^F = \sum_{1 \leq k \leq \ell} c_F^k \hat{\rho}_k^F + \sum_{1 \leq k_1 < k_2 \leq \ell} c_F^{k_1,k_2} \hat{\rho}_{k_1}^F \wedge \hat{\rho}_{k_2}^F + \dots$$

(finite sum).

But if $F_1 \subset F_2$, γ^{F_1} is the restriction of γ^{F_2} to $\underline{u}(F_1)$. It follows clearly from the finite dimensional case that the constants c 's do not depend on F for dimF large enough. Hence $J^*(\underline{u}_o)$ is an exterior algebra generated by the $\hat{\rho}_k$'s .

Step three. The primitivity and the unicity of the generators $\hat{\rho}_k$'s can be proved similarly, using finite dimensional subalgebras of $\underline{u}_o$. ∎

Let now $p \in \bar{\mathbb{R}}$ with $1 \leq p \leq \infty$, and let $\underline{u}_p$ denote for short the Lie algebra of compact operators $\underline{u}(\mathcal{H} ; C_p)$. For each $k \in N^*$ with $2k-1 \geq p$, $\hat{\rho}_k$ extends uniquely to a continuous $(2k-1)$-cochain on $\underline{u}_p$ which will be denoted by the same letter.

Corollary. The cochains $(\hat{\rho}_k)_{\substack{k \in N^* \\ 2k-1 \geq p}}$ are primitive cocycles on $\underline{u}_p$. The algebra $J^*_c(\underline{u}_p)$ is an exterior algebra generated by $(\hat{\rho}_k)_{k \in N^*,\ 2k-1 \geq p}$; moreover, these primitive generators are unique up to multiplication by non-zero real numbers. In particular $J^*_c(u_\infty) = J^o(u_\infty) = \mathbb{R}$.

Proof : clear, as $\underline{u}_o$ is dense in $\underline{u}_p$ and as those $\hat{\rho}_k$'s which are defined must be continuous. ∎

Remarks.

i) Proposition 1A still holds for the algebra $\underline{su}(\mathcal{H} ; C_o)$ and the set of generators $(\hat{\rho}_k)_{k \in N^*, k \geq 2}$; the corollary still holds for $\underline{su}(\mathcal{H} ; C_1)$ in the same way.

ii) The same description as above is also valid for the complex Lie algebras $\underline{gl}(\mathcal{H} ; C_p)$, $p \in \bar{\mathbb{R}}$ with $p = 0$ or $1 \leq p \leq \infty$, and $\underline{sl}(\mathcal{H} ; C_p)$, $p = 0$ or $p = 1$.

Let now $\mathcal{H}_{\mathbb{R}}$ be a real Hilbert space, which can be considered as the set of fixed points of a conjugation in $\mathcal{H}$. Let $\underline{so}_o$ be the Lie algebra of finite rank operators $\underline{o}(\mathcal{H}_{\mathbb{R}} ; C_o)$. It is easy to check that the restriction of q_k to $\underline{so}_o$ vanishes when $k \equiv 1 \pmod 4$. For each $k \in N^*$, let $\hat{\sigma}_{2k}$ be the restriction of q_{4k-1} to $\underline{so}_o$.

Proposition 1B. The cochains $(\hat{\sigma}_{2k})_{k \in N^*}$ are primitive cocycles on $\underline{so}_o$. The algebra $J^*(\underline{so}_o)$ is an exterior algebra generated by $(\hat{\sigma}_{2k})_{k \in N^*}$; moreover, these primitive generators are unique up to multiplication by non-zero real numbers.

Proof : as for proposition 1A. ∎

Let $p \in \bar{\mathbb{R}}$ with $1 \leq p \leq \infty$, and let $\underline{o}_p$ denote for short the Lie algebra of compact operators $\underline{o}(\mathcal{H}_{\mathbb{R}}; C_p)$ ($\underline{o}_p$ is also denoted by $\underline{so}_1$ when $p = 1$). For each $k \in N^*$ with $4k-1 \geq p$, $\hat{\sigma}_{2k}$ extends uniquely to a continuous $(4k-1)$-cochain on $\underline{o}_p$ which will be denoted by the same letter.

Corollary. Proposition 1B holds if $\underline{so}_o$ is replaced by $\underline{o}_p$, $J^*(\underline{so}_o)$ by $J^*_c(\underline{o}_p)$, and $(\hat{\sigma}_{2k})_{k \in N^*}$ by $(\hat{\sigma}_{2k})_{k \in N^*}$, $4k-1 \geq p$. In particular $J^*_c(\underline{o}_\infty) = \mathbb{R}$.

Let now $\mathcal{H}_{\mathbb{Q}}$ be a quaternionic Hilbert space, which can be considered as $\mathcal{H}$ furnished with an anticonjugation. Let $\underline{sp}_o$ be the Lie algebra of finite rank operators $\underline{sp}(\mathcal{H}_{\mathbb{Q}}; C_o)$. It is again easy to check that the restriction of q_k to $\underline{sp}_o$ vanishes when $k \equiv 1 \pmod 4$. For each $k \in N^*$, let $\hat{\tau}_{2k}$ be the restriction of $2q_{4k-1}$ to $\underline{sp}_o$.

Proposition 1C. The cochains $(\hat{\tau}_{2k})_{k \in N^*}$ are primitive cocycles on $\underline{sp}_o$. The algebra $J^*(\underline{sp}_o)$ is an exterior algebra generated by $(\hat{\tau}_{2k})_{k \in N^*}$; moreover, these primitive generators are unique up to multiplication by non-zero real numbers.

Proof : as for proposition 1A. ∎

Let $p \in \bar{\mathbb{R}}$ with $1 \leqslant p \leqslant \infty$, and let $\underline{sp}_p$ denote for short the Lie algebra of compact operators $\underline{sp}(\mathcal{H}_{\mathbb{Q}}; C_p)$. For each $k \in N^*$ with $4k-1 \geqslant p$, $\hat{\tau}_{2k}$ extends uniquely to a continuous $(4k-1)$-cochain on $\underline{sp}_p$ which will be denoted by the same letter.

Corollary. Proposition 1C holds if $\underline{sp}_o$ is replaced by $\underline{sp}_p$, $J^*(\underline{sp}_o)$ by $J^*_c(\underline{sp}_p)$, and $(\hat{\tau}_{2k})_{k \in N^*}$ by $(\hat{\tau}_{2k})_{k \in N^*}$, $4k-1 \geqslant p$. In particular $J^*_c(\underline{sp}_\infty) = \mathbb{R}$.

If $\mathbb{K}$ is now one of $\mathbb{R}$, $\mathbb{C}$, $\mathbb{Q}$, let $Hilb^+(\mathcal{H}_{\mathbb{K}}; C_p)$ denote the connected component of the group defined section III.1, and $\underline{hilb}(\mathcal{H}_{\mathbb{K}}; C_p)$ its Lie algebra.

Proposition 2. The real cohomology algebra of the Banach-Lie group $Hilb^+(\mathcal{H}_{\mathbb{K}}; C_p)$ is isomorphic to the algebra of continuous invariant cochains on the Banach-Lie algebra $\underline{hilb}(\mathcal{H}_{\mathbb{K}}; C_p)$ if and only if $p = 1$.

Proof : Proposition 2 follows by comparison between the corollary to proposition II.16 and the well-known results about the cohomology rings of the finite dimensional classical Lie groups on the one hand, and proposition IV.1 on the other hand. ∎

Remark. In the Riemannian case $Hilb^+(\mathcal{H}_{\mathbb{K}}; C_2)$, the explicit form of the generators of $J^*_c(\underline{hilb}(\mathcal{H}_{\mathbb{K}}; C_2))$ provides the harmonic differential forms explicitly on the group; a tiny part of Hodge's theory can be recuperated in this way. The unitary case $U(\mathcal{H}_{\mathbb{C}}; C_2)$ is remarkable in so far as there is no harmonic form to generate the first cohomology group.

IV.2. - The cohomology algebra $H_c^*(\underline{g})$

Let $\underline{g}$ be a Banach-Lie algebra. From the usual conventions, it follows that $H_c^0(\underline{g})$ is identified with the base field. It is an immediate consequence of the definitions that $H_c^1(\underline{g})$ is isomorphic to the topological dual of the Banach space $\underline{g}/\overline{[\underline{g},\underline{g}]}$. Hence, as long as scalar-valued cohomology is concerned, real problems start with $H_c^2(\underline{g})$. In this section, $H_c^2(\underline{g})$ is computed when $\underline{g}$ is an infinite dimensional classical complex Lie algebra of compact operators.

Let $\mathcal{H}$ be an infinite dimensional separable complex Hilbert space. Let $p \in \bar{\mathbb{R}}$ with $1 \leq p \leq \infty$, and let $\underline{a}$ be the Banach-Lie algebra $\underline{gl}(\mathcal{H}; C_p)$. Let $q \in \bar{\mathbb{R}}$ be defined by $\frac{1}{p} + \frac{1}{q} = 1$, and let $\underline{b}$ be the Banach-Lie algebra $\underline{gl}(\mathcal{H}; C_p)$ if $q < \infty$, and the Banach-Lie algebra $\underline{gl}(\mathcal{H}; L)$ if $q = \infty$. Consider as in section II.5 (the last remark) the duality

$$\langle\langle | \rangle\rangle \; : \; \begin{cases} \underline{a} \times \underline{b} \longrightarrow \mathbb{C} \\ (X,Y) \longmapsto \operatorname{trace}(XY^*) \end{cases} .$$

Let now ω be a 2-cochain on $\underline{a}$. According to the duality, ω defines a unique continuous linear operator $\Delta : \underline{a} \longrightarrow \underline{b}$ such that $\omega(X,Y) = \langle\langle X | \Delta(Y^*) \rangle\rangle$ for all $X,Y \in \underline{a}$; naturally, as ω is skew-symmetric, $\langle\langle X | \Delta(Y^*) \rangle\rangle = - \langle\langle Y | \Delta(X^*) \rangle\rangle$ for all $X,Y \in \underline{a}$. It follows directly from the definitions that ω is a cocycle [resp. a coboundary] if and only if Δ is a derivation [resp. an inner derivation]. Write $\operatorname{Der}(\underline{a},\underline{b})$ the space of all derivations from $\underline{a}$ to $\underline{b}$ and $\operatorname{Int}(\underline{a},\underline{b})$ the subspace of $\operatorname{Der}(\underline{a},\underline{b})$ containing those derivations Δ for which there exists $D \in \underline{b}$ with $\Delta(X) = [D,X]$ for all $X \in \underline{a}$. Hence $H_c^2(\underline{a}) = \operatorname{Der}(\underline{a},\underline{b}) / \operatorname{Int}(\underline{a},\underline{b})$ is explicitly known from section I.2.

<u>Proposition 3A</u>. Let p, q and $\underline{a}$ be as above. Then the vector space $H^2_c(\underline{a})$ is isomorphic to :

$$\begin{cases} \{0\} & \text{when } p = \infty \\ C_r(\mathcal{H})/C_q(\mathcal{H}), \text{ with } \frac{1}{r} + \frac{1}{p} = \frac{1}{q} & \text{when } 2<p\leq\infty \\ (L(\mathcal{H})/\mathbb{C}\mathrm{id}_{\mathcal{H}})/C_q(\mathcal{H}) & \text{when } 1<p\leq 2 \\ \{0\} & \text{when } p = 1 \end{cases}.$$

<u>Proof</u> : it is an immediate consequence of the considerations which precede the proposition and of the following lemma. ∎

<u>Lemma</u>. Let D be a continuous operator on $\mathcal{H}$. Let $p \in \bar{\mathbb{R}}$ with $2<p\leq\infty$, and let $q,r \in \bar{\mathbb{R}}$ be such that $\frac{1}{p} + \frac{1}{q} = 1$ and $\frac{1}{r} + \frac{1}{p} = \frac{1}{q}$. Suppose that $[D,X] \in C_q(\mathcal{H})$ for all $X \in C_p(\mathcal{H})$. Then $D \in C_r(\mathcal{H})$.

<u>Proof</u>. Let $\mathcal{I} = \{Y \in L(\mathcal{H}) \mid [Y,X] \in C_q(\mathcal{H})$ for all $X \in C_p(\mathcal{H})\}$; then $\mathcal{I}$ is clearly a non trivial Lie ideal of $L(\mathcal{H})$ which contains D. Hence (see proposition II.1A), D must be compact (up to a scalar multiple of the identity of $\mathcal{H}$). It is then sufficient to prove the lemma when D is a compact positive operator.

Let $(e_n)_{n \in N}$ be an orthonormal basis of $\mathcal{H}$ which diagonalises D and let $\lambda = (\lambda_n)_{n \in N}$ be a decreasing sequence of positive real numbers such that $D = \sum_{n\in N} \lambda_n e_n \otimes \overline{e_n}$. We want to show that $\lambda \in \ell^r$.

Let f be a map from N into itself such that $\lambda_{f(n)} \leq \frac{1}{2}\lambda_n$ for all $n \in N$. For each sequence of real numbers $\alpha = (\alpha_n)_{n \in N} \in \ell^p$, let X_α be the operator $\sum_{n\in N} \alpha_n e_n \otimes \overline{e_{f(n)}}$, which is in $C_p(\mathcal{H})$.
Then $[D,X_\alpha] = \sum_{n\in N} \alpha_n(\lambda_n - \lambda_{f(n)}) e_n \otimes \overline{e_{f(n)}}$ must be in $C_q(\mathcal{H})$,

hence the sequence of real numbers $\left(\alpha_n(\lambda_n - \lambda_{f(n)})\right)_{n \in N}$ must be in ℓ^q, and by the choice of f so must be $(\alpha_n \lambda_n)_{n \in N}$. This being true for all $\alpha \in \ell^p$, the sequence λ is in ℓ^r, whence the lemma. ∎

Conjectures

i) $H^*_c(\underline{gl}(\mathcal{H}; C_1))$ is isomorphic to $J^*_c(\underline{gl}(\mathcal{H}; C_1))$.

ii) $H^*_c(\underline{gl}(\mathcal{H}; C_2))$ is generated by the canonical image of $J^*_c(\underline{gl}(\mathcal{H}; C_2))$ and by $H^2_c(\underline{gl}(\mathcal{H}; C_2)) \approx (L(\mathcal{H})/\mathbb{C}id_{\mathcal{H}})/C_2(\mathcal{H})$.

iii) $H^*_c(\underline{gl}(\mathcal{H}; C_\infty)) = H^0(\underline{gl}(\mathcal{H}; C_\infty)) = \mathbb{C}$.

Remarks.

i) If $p \neq 1$, the real cohomology algebra of the Banach-Lie group $U(\mathcal{H}; C_p)$ is not isomorphic to the scalar cohomology $H^*_c(\underline{u}(\mathcal{H}; C_p))$ of its Lie algebra. Indeed, the first Betti number of $U(\mathcal{H}; C_p)$ is then equal to $+1$, while $H^1_c(\underline{u}(\mathcal{H}; C_p)) = \{0\}$; moreover, if $p \neq \infty$, the second Betti number of $U(\mathcal{H}; C_p)$ vanishes, while $\dim H^2_c(\underline{u}(\mathcal{H}; C_p)) = \infty$.

ii) If the conjecture i) above was true, then the real cohomology of $U(\mathcal{H}; C_1)$ would be isomorphic to $H^*_c(\underline{u}(\mathcal{H}; C_1))$; similarly for the group $SU(\mathcal{H}; C_1)$.

iii) In the L^*-case $p = 2$, the canonical image of $J^*_c(\underline{gl}(\mathcal{H}; C_2))$ is a proper subalgebra $\mathcal{F}$ in $H^*_c(\underline{gl}(\mathcal{H}; C_2))$ which has the following property: let $\underline{s}$ be a classical simple complex Lie algebra of finite dimension which is a subalgebra of $\underline{gl}(\mathcal{H}; C_2)$; then the restrictions of $\mathcal{F}$ and of $H^*_c(\underline{gl}(\mathcal{H}; C_2))$ to $\underline{s}$ are equal. $\mathcal{F}$ is in this sense a kind of "finite approximation" for $H^*_c(\underline{gl}(\mathcal{H}; C_2))$.

iv) It is known that the first and second scalar cohomology spaces of the associative C^*-algebra $C(\mathcal{H})$ do vanish; see Guichardet [69].

v) **Proposition II.10 can also be** expressed in a cohomology formulation about $H^1(\underline{g},\underline{g})$ ($\underline{g}$-valued cohomology of $\underline{g}$).

vi) Any statement and conjecture about $\underline{gl}(\mathcal{H}; C_1)$ has an immediate and equivalent counterpart for $\underline{sl}(\mathcal{H}; C_1)$ which we will not write down explicitly.

Proposition 3B. Let $J_{\mathbb{R}}$ be a conjugation of $\mathcal{H}$; let $p \in \bar{\mathbb{R}}$ with $1 \leqslant p \leqslant \infty$ and let $q \in \bar{\mathbb{R}}$ be defined by $\frac{1}{p} + \frac{1}{q} = 1$. Consider the Banach-Lie algebra $\underline{a} = \underline{o}(\mathcal{H}, J_{\mathbb{R}}; C_p)$. Then the vector space $H^2_c(\underline{a})$ is isomorphic to

$\{0\}$ when $p = \infty$

$\underline{o}(\mathcal{H}, J_{\mathbb{R}}; C_r) / \underline{o}(\mathcal{H}, J_{\mathbb{R}}; C_q)$, with $\frac{1}{r} + \frac{1}{p} = \frac{1}{q}$ when $2 < p \leqslant \infty$

$\underline{o}(\mathcal{H}, J_{\mathbb{R}}; L) / \underline{o}(\mathcal{H}, J_{\mathbb{R}}; C_q)$ when $1 < p \leqslant 2$

$\{0\}$ when $p = 1$.

Proposition 3C. It is obtained by replacing $J_{\mathbb{R}}$ by $J_{\mathbb{Q}}$ and $\underline{o}$ by $\underline{sp}$ in proposition 3B.

Proofs : as for proposition 3A. ∎

Remarks : as those following proposition 3A.

IV. 3. - The algebra $I(G)$ of invariant polynomials

Let $\mathfrak{g}$ be a Lie algebra over $\mathbb{K}$ ($\mathbb{K}$ is $\mathbb{R}$ or $\mathbb{C}$). We will denote by $S_c(\mathfrak{g}) = \bigoplus_{k \in N} S_c^k(\mathfrak{g})$ the Z-graded commutative $\mathbb{K}$-algebra with unit defined by the continuous symmetric multilinear maps from $\mathfrak{g}$ to $\mathbb{K}$, and by $P_c(\mathfrak{g})$ that defined by the continuous polynomial maps on $\mathfrak{g}$. It is well-known that $S_c(\mathfrak{g})$ and $P_c(\mathfrak{g})$ are naturally isomorphic (see for example Douady [48] section 1.1).

Suppose now that $\mathfrak{g}$ is the Lie algebra of a Banach-Lie group G. A function $F \in S_c^k(\mathfrak{g})$ is <u>invariant</u> by G if $F(gX_1, \ldots, gX_k) = F(X_1, \ldots, X_k)$ for all $X_1, \ldots, X_k \in \mathfrak{g}$ and for all $g \in G$, where gX denotes the result of the transform of X by g according to the adjoint action of G in $\mathfrak{g}$. Invariant functions define a subalgebra of $S_c(\mathfrak{g})$ which will be denoted by $S(G) = \bigoplus_{k \in N} S^k(G)$. Similarly, invariant polynomials define a subalgebra $I(G)$ of $P_c(\mathfrak{g})$; the natural isomorphism between $S_c(\mathfrak{g})$ and $P_c(\mathfrak{g})$ induces an isomorphism between $S(G)$ and $I(G)$. In this section, we want to describe $I(G)$ when G is one of the classical Banach-Lie group of compact operators $Hilb(\mathcal{H}_{\mathbb{K}}; C_p)$, with $\mathcal{H}_{\mathbb{K}}$ an infinite dimensional separable Hilbert space over $\mathbb{K}$.

Let first $\mathcal{H}$ be a complex space, let $p \in \bar{\mathbb{R}}$ with $1 \leq p \leq \infty$, and consider the group $U_p = U(\mathcal{H}; C_p)$. To any orthonormal basis $e = (e_n)_{n \in N}$ in $\mathcal{H}$ corresponds a maximal torus T in U_p, and the Lie algebra of T is isomorphic to the space of real sequences ℓ^p (when $p < \infty$) or c_o (when $p = \infty$). The inclusion $T \longrightarrow U_p$ induces a morphism $I(U_p) \longrightarrow I(T)$. Hence, functions in $I(U_p)$ can be looked at as continuous polynomial functions on ℓ^p (or c_o if $p = \infty$) which are invariant with respect to the infinite symmetric group W_A (see proposition I.7A). By following the same method as in section IV.1, and by using the standard Newton theorem on elementary symmetric functions, it is easy to compute explicitly $I(U_p)$. We will give now the result and leave the easy checking to the reader (see as well [83]).

Let $k \in N^*$, $k \geq p$. Let ρ_k be the continuous polynomial function defined by $\rho_k(X) = \text{trace}\,\{(-\sqrt{-1}\,X)^k\}$ for all $X \in \underline{u}(\mathcal{H}; C_p)$.

<u>Proposition 4A</u>. The algebra $I(U_p)$ is a polynomial algebra generated by the functions $(\rho_k)_{k \in N^*, k \geq p}$. In particular $I(U_\infty) = I^0(U_\infty) = \mathbb{R}$.

<u>Remarks</u>.

i) The generators $(\rho_k)_{k \in N^*, k \geq p}$ are not unique. Indeed, if $p = 1$, other systems of generators are provided by the invariant Chern functions and the dual invariant Chern functions; see Murakami [125].

ii) Propositions 4B and 4C are analogous to proposition 4A and are left to the reader. Similarly for $I(SU(\mathcal{H}; C_1))$.

$Hilb^+(\mathcal{H}_{\mathbb{K}}; C_p)$ being again as in pages III.4 and IV.6, one has :

<u>Proposition 5</u>. The real cohomology algebra of the classifying space of the Banach-Lie group $Hilb^+(\mathcal{H}_{\mathbb{K}}; C_p)$ is isomorphic to the algebra $I(Hilb^+(\mathcal{H}_{\mathbb{K}}; C_p))$ if and only if $p = 1$.

<u>Proof</u> : proposition 5 follows immediately from propositions III.1 and IV.4. ∎

Projects.

1. - Make explicit the map T from $I(\mathrm{Hilb}^{+}(\mathcal{H}_{\mathbb{K}};\ C_p))$ to $J_c^{*}(\underline{\mathrm{hilb}}(\mathcal{H}_{\mathbb{K}};\ C_p))$ which sends ρ_k to $\hat{\rho}_k$ for all $k \in N$, $k \geqslant p$ (see the bottom of page IV.1). The image of T will contain the canonical set of generators in $J_c^{*}(\underline{\mathrm{hilb}}(\mathcal{H}_{\mathbb{K}};\ C_p))$ if and only if $1 \leqslant p \leqslant 2$ or $p = \infty$, as it is immediately seen from the corollary to proposition IV.1 and from proposition IV.4.

2. - Study explicitly the Weil algebras (see Cartan [31], [32]) of the classical complex Banach-Lie algebras of compact operators, especially in the case of nuclear operators ($p = 1$) . This project would best follow some work on conjectures i) and ii) section IV.2.

BIBLIOGRAPHY

1. - H.R. Alagia : Conjugate classes of Cartan subalgebras of real simple L*-algebras. Thesis, Washington University, August 1971.

2. - W. Ambrose : Structure theorems for a special class of Banach algebras. Trans. AMS 57 (1945) 364-386. MR 7-126.

3. - B.H. Arnold : Ring of operators on vector spaces. Ann. of Math. 45 (1944) 24-49. MR 5-147.

4. - V. Arnold : Sur la courbure de Riemann des groupes de difféomorphismes. C.R. Acad. Sci. Paris 260 (1965) 5668-5671. MR 31 +2692.

5. - --- : Sur la géométrie différentielle des groupes de Lie de dimension infinie et ses applications à l'hydrodynamique des fluides parfaits. Ann. Inst. Fourier (Grenoble) 16 (1966) 319-361. MR 34 +1956.

6. - M.F. Atiyah, R. Bott and A. Shapiro : Clifford modules. Topology vol. 3 suppl. 1 (1964) 3-38. MR 29 +5250.

7. - M.F. Atiyah and I.M. Singer : Index theory for skew-adjoint Fredholm operators. Publ. Math. IHES 37 (1969).

8. - V.K. Balachandran : The Weyl group of an L*-algebra. Math. Ann. 154 (1964) 157-165. MR 28 +4374.

9. - --- : Regular elements of L*-algebras. Math. Z. 93 (1966) 161-163. MR 33 +2689.

10.- --- : On the uniqueness of the inner product topology in a semi-simple L*-algebra. Topology 7 (1968) 305-309. MR 37 +4137.

11.- --- : Simple systems of roots in L*-algebras. Trans. AMS 130 (1968) 513-524. MR 38 +199.

12.---- : Simple L-algebras of classical type. Math.Annalen 180 (1969) 205-219. MR 39 +4684.

13.- --- : Fixed-point-free automorphisms in L*-algebras. Arch. der Math. 21 (1970) 386-389. MR 42 +7721.

14.- --- : Automorphisms of L*-algebras. Tôhoku J. Math. 22 (1970) 163-173. MR 42 +7722.

15.---- : Real L-algebras. To be published.

16. - --- and P.R. Parthasarathy : Cartan subalgebras of an L*-algebra. Math. Ann. 166 (1966) 300-301. MR 34 +3348.

17. - --- and P.S. Rema : Uniqueness of the norm topology in certain Banach-Jordan algebras. Publ. Ramanujan Inst. 1 (1968/9) 283-289. MR 42 +5246.

18. - V. Bargman : Note on Wigner's theorem on symmetry operations. J. of Math. Phys. 5 (1964) 862-868. MR 29 +1917.

19. - Bateman's Manuscript Project : Higher transcendental functions, vol. II. McGraw Hill 1953.

20. - G. Birkhoff : Analytical groups. Trans. AMS 43 (1938) 61-101.

21. - R.J. Blattner : Automorphic group representations. Pacific J. Math. 8 (1958) 665-677. MR 21 +2191.

*22. - A. Borel : Topics in the homology theory of fibre bundles. Lecture Notes in Mathematics 36. Springer 1967.

23. - --- and G.D. Mostow : On semi-simple automorphisms of Lie algebras. Ann. of Math. 61 (1955) 389-405. MR 16-897.

24. - N. Bourbaki : Algèbre, chapitre 9. Hermann 1959.

*25.- --- : Groupes et algèbres de Lie, chapitre 1. Hermann 1960.

26. - --- : Groupes et algèbres de Lie, chapitres IV-VI. Hermann 1968.

27. - --- : Variétés différentielles et analytiques. Fascicule de résultats (paragraphes 1 à 7). Hermann 1967.

28. - --- : Variétés différentielles et analytiques. Fascicule de résultats (paragraphes 8 à 15). Hermann 1971.

29. - R.H. Bruck and E. Kleinfeld : The structure of alternative division rings. Proc. AMS 2 (1951) 878-890. MR 13-526.

*30.- J.W. Calkin : Two-sided ideals and congruences in the ring of bounded operators in Hilbert space. Ann. of Math. 42 (1941) 839-873. MR 3-208.

*31. - H. Cartan : Notions d'algèbre différentielle; application aux groupes de Lie et aux variétés où opère un groupe de Lie. Colloque de topologie (espaces fibrés), Bruxelles, 1950, pages 15-27. MR 13-107.

*32. - ---: La transgression dans un groupe de Lie et dans un espace fibré principal. Ibid., pages 57-71. MR 13-107.

33.- --- and S. Eilenberg : Homological algebra. Princeton University Press 1956.

34.- J. Cerf : Topologie de certains espaces de plongements. Bull. Soc. Math. France 89 (1961) 227-330. MR 25 +3543.

35.- S. Chern : Topics in differential geometry. The Institute for Advanced Study, Princeton 1951.

36.- C. Chevalley : Theory of Lie groups. Princeton University Press 1946.

37.- --- : Théorie des groupes de Lie, groupes algébriques, théorèmes généraux sur les algèbres de Lie. Hermann 1968.

38.- --- and S. Eilenberg : cohomology theory of Lie groups and Lie algebras. Trans. AMS 63 (1948) 85-124. MR 9-567.

39.- J.S. Clowes and K.A. Hirsch : Simple groups of infinite matrices. Math. Z. 58 (1953) 1-3. MR 14-1060, 1279.

40.- Ja. L. Dalec'kiĭ : Infinite dimensional elliptic operators and parabolic equations connected with them. Russian Math. Surveys 29 (1970) n°4, 1-53.

41.- --- and Ja. I. Snaĭderman : Diffusion and quasiinvariant measures on infinite-dimensional Lie groups. Funkcional Anal. i Prilozen 3 (1969) n°2, 88-90. MR 40 +2138.

42.- J. Delsarte : Les groupes de transformations linéaires dans l'espace de Hilbert. Mem. des Sc. Math. LVII. Gauthiers-Villars 1932.

43.- J. Dieudonné : Die Lieschen Gruppen in der modernen Mathematik (see the discussion with K. Bleuler). Arbeitsgem. Forsch. Nordrhein-Westfalen Natur. Ingen. Gesellschaftwiss. Heft 133 (1964) 7-30. MR 30 +4853.

44.- --- : Fondements de l'analyse moderne. Gauthiers-Villars 1963.

45.- --- : Eléments d'analyse, 2. Gauthiers-Villars 1968.

46.- J. Dixmier : Les algèbres d'opérateurs dans l'espace hilbertien (algèbres de von Neumann), deuxième édition. Gauthier-Villars 1969.

47.- --- : Les C*-algèbres et leurs représentations, deuxième édition. Gauthier-Villars 1969.

48.- A. Douady : Le problème des modules pour les sous-espaces analytiques compacts d'un espace analytique donné. Ann. Inst. Fourier (Grenoble) 16 (1966) 1-95. MR 34 +2940.

49.- --- et M. Lazard : Espaces fibrés en algèbres de Lie et en groupes. Invent. Math. 1 (1966) 133-151. MR 33 +5787.

50.- N. Dunford and J.T. Schwartz : Linear operators II. Interscience 1963.

51.- E.B. Dynkin : Normed Lie algebras and analytic groups. AMS Translations 97 (1953). MR 15-282.

52.- --- : Topological characteristics of homomorphisms of compact Lie groups. AMS Translations (2) 12 (1959) 301-342. MR 16-673.

53.- J. Eells, Jr. : A class of smooth bundles over a manifold. Pacific J. Math. 10 (1960) 525-538. MR 22 +11405.

*54.- --- : A setting for global analysis. Bull. AMS 72 (1966) 751-807. MR 34 +3590.

55.- --- : Fredholm structures. Proc. Symp. Non-linear Functional Analysis, Chicago, 1968. Proc. Symp. Pure Math. 18 AMS 1970.

56.- --- and K.D. Elworthy : On Fredholm manifolds. Congrès international des mathématiciens, Nice, 1970. Actes, 2 (1971) 215-219.

57.- --- and K.D. Elworthy : Wiener integration on certain manifolds. University of Warwick, 1970.

58.- --- and K.D. Elworthy : Seminar on Wiener integration. University of Warwick, summer term, 1971.

59.- --- : Integration on Banach manifolds. To appear in the Proceedings of the Thirteenth Biennial Seminar of the Canadian Mathematical Congress (August-September 1971).

60.- M. Eidelheit : On isomorphisms of rings of linear operators. Studia Math. 9 (1940) 97-105. MR 2-224, 3-51, 7-620.

61.- K.D. Elworthy : Fredholm maps and $Gl_c\langle E\rangle$ structures. Bull. AMS 74 (1968) 582-587. MR 36 +7160.

62.- --- and A.J. Tromba : Differential structures and Fredholm maps on Banach manifolds. Global Analysis, Berkeley, July 1968. Proc. Sympos. Pure Math., XV (1970) 45-94. MR 41 +9299.

63.- L. Eulero : Introductio in analysin infinitorum, tomi primi. M.-M. Bousquet & Socios, Lausannae MDCCXLVIII.

*64.- K. Geba : On the homotopy groups of $CL_c(E)$. Bull. Acad. Polon. Sci. Sér. Sci. Math. Astronom. Phys. 16 (1968) 699-702. MR 38 +5249.

65.- I.C. Gohberg and M.G. Krein : Introduction to the theory of linear non-self-adjoint operators. Translations of mathematical monographs, vol. 18, AMS 1969. MR 39 +7447.

66.- S.J. Greenfield and N.R. Wallach : The Hilbert ball and bi-ball are holomorphically inequivalent. Bull. AMS 77 (1971) 261-263.

67.- --- : Automorphism groups of bounded domains in Banach spaces. To be published.

68.- L. Gross : Abstract Wiener spaces and infinite dimensional potential theory. Lectures in modern analysis and applications II pp. 84-116. Lecture Notes in Mathematics 140. Springer 1970. MR 42 +457.

69.- A. Guichardet : Sur l'homologie et la cohomologie des algèbres de Banach. C.R. Acad. Sci. Paris, Sér. A, 262 (1966) 38-41. MR 32 +8199.

70.- --- : Produits tensoriels infinis et représentations des relations d'anticommutation. Ann. Scient. Ec. Norm. Sup. (3) 83 (1966) 1-52. MR 34 +4932.

71.- A. Haefliger : Sur l'extension du groupe structural d'un espace fibré. C.R. Acad. Sci. Paris 243 (1956) 558-560. MR 18-920.

72.- V.L. Hansen : Some theorems on direct limits of expanding sequences of manifolds. Math. Scand. 29 (1971) 5-36.

73.- L.A. Harris : Schwarz's lemma in normed linear spaces. Proc. Nat. Acad. Sci. U.S.A. 62 (1969) 1014-1017.

74.- --- : Schwarz lemmas for Jordan algebras of operators. To be published.

75.- P. de la Harpe : Notes on Hilbert manifolds. University of Warwick, June 1970.

76.- --- : Classification of simple real L*-algebras. University of Warwick, July 1970.

77.- --- : Classification des L*-algèbres semi-simples réelles séparables. C.R. Acad. Sci. Paris, Sér. A, 272 (1971) 1559-1561.

78.- --- : L*-algèbres simples et algèbres de Lie classiques d'opérateurs dans l'espace hilbertien. C.R. Acad. Sci. Paris, Sér. A, 274 (1972) 1096-1098.

79.- --- : Groupe spinoriel associé à un espace de Hilbert. C.R. Acad. Sci. Paris, Sér. A, 273 (1971) 305-307.

80.- --- : Dérivations des L*-algèbres semi-simples complexes et des algèbres de Lie classiques d'opérateurs compacts. C.R. Acad. Sci. Paris, Sér. A, 273 (1971) 806-809.

81.- --- : Calcul des éléments primitifs dans la cohomologie des algèbres de Lie classiques. Universités de Warwick et Lausanne, Septembre 1971.

82.- --- et R. Ramer : Cohomologie scalaire des algèbres de Lie classiques de type A d'opérateurs compacts. C.R. Acad. Sci. Paris, Sér. A, 273 (1971) 882-885.

83.- --- et R. Ramer : Polynômes invariant sur les algèbres de Lie banachiques complexes classiques d'opérateurs compacts dans l'espace hilbertien. C.R. Acad. Sci. Paris, Sér. A, 274 (1972) 824-827.

*84.- S. Helgason : Differential geometry and symmetric spaces. Academic Press 1962.

85.- I.N. Herstein : Lie and Jordan systems in simple rings with involution. Amer. J. Math. 78 (1956) 629-649. MR 18 +714.

*86.- --- : Lie and Jordan structures in simple associative rings. Bull. AMS 67 (1961) 517-531. MR 25 +3072.

87.- --- : On the Lie structure of an associative ring. J. Algebra 14 (1970) 561-571. MR 41 +270.

88.- F. Hirzebruch : Topological methods in algebraic geometry. Third edition, Springer 1966.

89.- L. Illusie : Contractibilité du groupe linéaire des espaces de Hilbert de dimension infinie. Séminaire Bourbaki 284 (1965).

90.- N. Jacobson : Lie algebras. Interscience 1962.

91.- B.E. Johnson : Norms of derivations on $\mathcal{L}(X)$. Pacific J. Math. 38 (1971) 465-469.

*92.- --- and A.M. Sinclair : Continuity of derivations and a problem of Kaplansky. Amer. J. Math. 90 (1968) 1067-1073. MR 39 +776.

93.- R.V. Kadison : Infinite unitary groups. Trans. AMS 72 (1952) 386-399. MR 14-16.

94.- --- : Infinite general linear groups. Trans. AMS 76 (1954) 66-91. MR 15-721.

95.- --- : On the general linear groups of infinite factors. Duke Math. J. 22 (1955) 119-122. MR 16-719.

96.- --- : Mappings of operator algebras. Congrès international des mathématiciens, Nice, 1970. Actes, 2 (1971) 389-393.

97.- I. Kaplansky : Dual rings. Ann. of Math. 49 (1948) 689-701. MR 10-7.

98.- M. Karoubi : Algèbres de Clifford et K-théorie. Ann. Scient. Ec. Norm. Sup. (4) 1 (1968) 161-270.

99.- B. Kostant : On the conjugacy of real Cartan subalgebras. I. Proc. Nat. Acad. Sci. U.S.A. 41 (1955) 967-970. MR 17-509.

*100.- J.L. Koszul : Homologie et cohomologie des algèbres de Lie. Bull. Soc. Math. France 78 (1950) 65-127. MR 12-120.

101.- N. Kuiper : The homotopy type of the unitary group of Hilbert space. Topology 3 (1965) 19-30. MR 31 +4034.

102.- --- : Variétés hilbertiennes : aspects géométriques. Suivi de deux textes de D. Burghelea. Séminaire de mathématiques supérieures, Montréal, été 1969.

*103.- S. Lang : Introduction aux variétés différentiables. Dunod 1967.

104.- D. Laugwitz : Über unendliche kontinuierliche Gruppen. I. Grundlagen der Theorie; Untergruppen. Math. Ann. 130 (1955) 337-350. MR 17-762.

105.- --- : Über unendliche kontinuierliche Gruppen. II. Strukturtheorie local-Banachscher Gruppen. Bayer Akad. Wiss. Math.-Nat. Kl. S.- 13. 1956 (1957) 262-286. MR 19-753.

106.- M. Lazard : Variétés différentiables. "Septième leçon" of a course to be published in book-form.

*107.- --- : Groupes différentiables. "Neuvième leçon" of the same.

108.- --- et J. Tits : Domaines d'injectivité de l'application exponentielle. Topology 4 (1965/6) 315-322. MR 32 +2518.

109.- J.A. Leslie : On a theorem of E. Cartan. Ann. Mat. Pura Appl. (4) 74 (1966) 173-177. MR 34 +5967.

110.- E.M. Levič : An example of a locally nilpotent group without torsion and without accessible elements. Sibirsk. Mat. Ž. 8 (1967) 717-719. MR 35 +6742.

111.- --- and A.I. Tokarenko : A remark on locally nilpotent torsion-free groups. Sibirsk. Mat. Ž. 11 (1970) 1406-1408. MR 42 +7777.

112.- N. Limic : Nilpotent locally convex Lie algebras and Lie fields structures. Institute for Advanced Study, Princeton. Apparently 1969.

113.- R.J. Loy : Uniqueness of the complete norm topology and continuity of derivations on Banach algebras. Tôhoku Math. J. 22 (1970) 371-378.

114.- J.H. McAlpin : Infinite dimensional manifolds and Morse theory. Thesis, Columbia University 1965.

115.- G.W. Mackey : Isomorphisms of normed linear spaces. Ann. of Math. 43 (1942) 244-260. MR 4-12.

116.- B. Maissen : Lie-Gruppen mit Banachräumen als Parameterräume. Acta Math. 108 (1962) 229-270.

117.- W.S. Martindale III : Lie derivations of primitive rings. Michigan Math. J. 11 (1964) 183-187. MR 29 +3511.

118.- --- : Lie isomorphisms of prime rings. Trans. AMS 142 (1969) 437-455. MR 40 +4308.

119.- A.D. Michal : Infinite dimensional differential metrics with constant curvature. Proc. Nat. Acad. Sci. U.S.A. 34 (1948) 17-21. MR 9-380.

120.- J. Milnor : Morse theory. Princeton University Press 1963.

121.- --- : Spin structures on manifolds. Enseignement Math. (2) 9 (1963) 198-203. MR 28 +622.

122.- --- : Remarks concerning spin manifolds. Differential and Combinatorial Topology (A Symposium in Honor of Marton Morse) pp. 55-62. Princeton University Press 1965. MR 31 +5208.

123.- B.S. Mityagini : The homotopy structure of the linear group of a Banach space. Russian Math. Surveys 25 (1970) 59-103.

124.- G.D. Mostow : Some new decomposition theorems for semi-simple groups. Mem. Amer. Math. Soc. 14 (1955) 31-54. MR 16-1087.

125.- S. Murakami : Algebraic study of fundamental characteristic classes of sphere bundles. Osaka Math. J. 8 (1956) 187-224.

126.- H. Omori and P. de la Harpe : About interactions between Banach-Lie groups and finite dimensional manifolds. To be published.

127.- --- : Opération de groupes de Lie banachiques sur les variétés différentielles de dimension finie. C.R. Acad. Sci. Paris, Sér. A, 273 (1971) 395-397.

128.- R. Ouzilou : Groupes classiques de Banach. C.R. Acad. Sci. Paris, Sér. A, 272 (1971) 1459-1461.

129.- R.S. Palais : On the homotopy type of certain groups of operators. Topology 3 (1965) 271-279. MR 30 +5315.

130.- --- : Homotopy theory of infinite dimensional manifolds. Topology 5 (1966) 1-16. MR 32 +6455.

131.- C. Pearcy and D. Topping : On commutators in ideals of compact operators. Michigan Math. J. 18 (1971) 247-252.

132.- A. Pietsch : Ideale von S_p-Operatoren in Banachräumen. Studia Math. 38 (1970) 59-69.

133.- --- :Adjungierte normierte Operatorenideale. Math. Nachr. 48 (1971) 189-211.

134.- C.R. Putnam : Commutation properties of Hilbert space operators and related topics. Springer 1967.

135.- --- and A. Wintner : The connectedness of the orthogonal group in Hilbert space. Proc. Nat. Acad. Sci. U.S.A. 37 (1951) 110-112. MR 13-10.

136.- --- and A. Wintner : The orthogonal group in Hilbert space. Amer. J. Math. 74 (1952) 52-78. MR 13-531.

137.- D. Quillen : The spectrum of an equivariant cohomology ring : I. Ann. of Math. 94 (1971) 549-572.

138.- C.E. Rickart : Isomorphic groups of linear transformations. Amer. J. Math. 72 (1950) 451-464. MR 11-729.

139.- --- : Isomorphic groups of linear transformations. II. Amer. J. Math. 73 (1951) 697-716. MR 13-532.

140.- --- : Isomorphisms of infinite dimensional analogues of the classical groups. Bull. AMS 57 (1951) 435-448. MR 13-532.

141.- --- : General theory of Banach algebras. Van Nostrand 1960.

142.- J.R. Ringrose : On the triangular representation of integral operators. Proc. London Math. Soc (2) 12 (1962) 385-399. MR 25 +3372.

143.- E.F. Robertson : A remark on the derived group of GL(R) . Bull. London Math. Soc. 1 (1969) 160-162. MR 40 +5745.

144.- A. Rosenberg : The structure of the infinite linear groups. Ann. of Math. 68 (1958) 278-294. MR 21 +1319.

145.- W. Rudin : Real and complex analysis. McGraw Hill 1966.

146.- S. Sakai : Derivations of simple C*-algebras. II. Bull. Soc. Math. France 99 (1971) 259-263.

147.- S. Saks and A. Zygmund : Analytic functions. Second edition. PWN, Warszawa 1965.

*148.- H. Samelson : Topology of Lie groups. Bull. AMS 58 (1952) 2-37. MR 13-533.

149.- P.P. Saworotnow : On a generalization of the notion of H*-algebra. Proc. AMS 8 (1957) 49-55. MR 19-47.

150.- --- : On the embedding of a right complemented algebra into Ambrose's H*-algebras. Proc. AMS 8 (1957) 56-62. MR 19-47.

*151.- R. Schatten : Norm ideals of completely continuous operators. Springer 1960.

152.- M. Schechter : Riesz operators and Fredholm perturbations. Bull. AMS 74 (1968) 1139-1144. MR 37 +6777.

*153.- J.R. Schue : Hilbert space methods in the theory of Lie algebras. Trans. AMS 95 (1960) 69-80. MR 22 +8352.

154.- --- : Cartan decomposition for L-algebras. Trans. Amer. Math. Soc. 98 (1961) 334-349. MR 24 +A3242.

155.- --- : The structure of hyperreducible triangular algebras. Proc. AMS 15 (1964) 766-772. MR 29 +3897.

156.- Séminaire Sophus Lie : Théorie des algèbres de Lie, topologie des groupes de Lie. Ecole Normale Supérieure, Paris,1955.

*157.- J.P. Serre : Algèbres de Lie semi-simples complexes. Benjamin 1966.

158.- --- : Corps locaux, deuxième édition. Hermann 1968.

159.- D. Shale : Linear symmetries of free boson fields. Trans. AMS 103 (1962) 149-167. MR 25 +956.

160.- --- and W.F. Stinespring : States of the Clifford algebra. Ann. of Math. 80 (1964) 365-381. MR 29 +3160.

161.- --- and W.F. Stinespring : Spinor representations of infinite orthogonal groups. J. Math. Mech. 14 (1965) 315-322. MR 30 +3377.

162.- L.A. Simonjan : Two radicals of Lie algebras. Dokl. Akad. Nauk SSSR 157 (1964) 281-283. MR 29 +1235.

163.- J. Sławny : Representations of canonical anticommutation relations and implementability of canonical transformations. Thesis, The Weizmann Institute of Science, Rehovot 1969.

164.- --- : Same title. Commun. Math. Phys. 22 (1971) 104-114.

165.- M.F. Smiley : Right H*-algebras. Proc. AMS 4 (1953) 1-4. MR 14-660.

166.- --- : Real Hilbert algebras with identity. Proc. AMS 16 (1965) 440-441. MR 31 +591.

167.- J.G. Stampfli : The norm of a derivation. Pacific J. Math. 33 (1970) 737-747.

168.- I.N. Stewart : Subideals of Lie algebras. Thesis, University of Warwick, 1969.

169.- --- : Lie algebras. Lecture Notes in Mathematics 127. Springer 1970.

170.- --- : An algebraic treatment of Mal'cev's theorems concerning nilpotent Lie groups and their Lie algebras. Compositio Math. 22 (1970) 289-312.

171.- H. Sunouchy : Infinite Lie rings. Tôhoku Math. J. (2) 8 (1956) 291-307. MR 21 +75.

172.- A.S. Svarc : The homotopic topology of Banach spaces. Dokl. 5 (1964) 57-59.

173.- S. Swierczkowski : The path-functor on Banach-Lie algebras. Proc. Kon. Ned. Akad. v. Wetensch., Amsterdam, 74 (1971) 235-239.

174.- Unsain : Classification of the simple separable real L-algebras. Thesis, University of California, Berkeley 1970.

175.- --- : Same title. Bull. AMS 77 (1971) 462-466.

176.- --- : Cartan-Ambrose theorem for infinite dimensional Riemannian manifolds. University of Maryland, September 1971.

177.- W.T. VanEst and Th.J. Korthagen : Nonenlargible Lie algebras. Nederl. Akad. Wetensch. Proc. Ser. A 67 = Indag. Math. 26 (1964) 15-31. MR 28 +4061.

178.- F.H. Vasilescu : Normed Lie algebras. Queen's University, Kingston, Preprint 21 (1971).

179.- --- : Positive forms on Lie algebras with involution. Queen's University, Kingston, Preprint 40 (1971).

180.- --- : On Lie's theorem in operators algebras. Queen's University, Kingston, Preprint 47 (1971).

181.- B.B. Venkov : Cohomology algebras for some classifying spaces. Dokl. Akad. Nauk SSSR 127 (1959) 943-944. MR 21 +7500.

182.- A. Weinstein : Symplectic structures on Banach manifolds. Bull. AMS 75 (1969) 1040-1041.

183.- --- : Symplectic manifolds and their Lagrangian submanifolds. To be published.

184.- E.T. Whithaker and G.N. Watson : A course of modern analysis, fourth edition. Cambridge University Press 1963.

185.- J.A. Wolf : Spaces of constant curvature. McGraw Hill 1967.

186.- F.B. Wright : Absolute-valued algebras. Proc. Nat. Acad. Sci. U.S.A. 39 (1953) 330-332.

187.- S. Yamaguchi : On simple Lie groups of infinite dimension. I. Mem. Faculty Sci. Kyushu Univ., Ser. A, Mathematics, 18 (1964) 1-32.

188.- --- : On simple Lie groups of infinite dimension. II. Same journal 18 (1964) 33-43.

189.- --- : Note on certain simple Lie algebras of infinite dimension. Same journal 18 (1964) 44-49.

190.- --- : Correction to "on simple Lie groups of infinite dimension II". Same journal 19 (1965) 105-107.

191.- P. Zvengrowski : Canonical vector fields on spheres. Commentarii 43 (1968) 341-347.

192.- C.R. Miers : Lie isomorphisms of factors. Trans. AMS 147 (1970) 55-63.

193.- --- : Lie homomorphisms of operator algebras. Pacific J. Math. 38 (1971) 717-735.

194.- N. Moulis : Structures de Fredholm sur les variétés hilbertiennes. Forthcoming mémoire.

195.- R. Ouzilou : Groupes et algèbres de Lie banachiques. Thèse, Université de Lyon I, 1972.

196.- F.H. Vasilescu : Radical d'une algèbre de Lie de dimension infinite. C.R. Acad. Sci. Paris, Sér. A, 274 (1972) 536-538.

197.- M.I. Vishik : The parametrix of elliptic operators with infinitely many independent variables. Russian Math. Surveys 26 n°2 (1971) 91-112.

198.- A. Jarry : Historique futur des groupes de Palotin-Lie. Forthcoming.

The last seven references came to the knowledge of the author after completion of the present work.

The sign * indicates a reference of more direct relevance.

Added in proof

A construction similar to that conjectured in project 8.2 (page II.35) has been carried out in "The Clifford algebra and the spinor group of a Hilbert space" (to be published). The universal covering Spin($\mathcal{H}$; C_1) of $O^+(\mathcal{H}; C_1)$ is described explicitely in terms of the C*-Clifford algebra of the quadratic form $x \mapsto -|x|^2$ on $\mathcal{H}$.

In the terminology of quantum physics (see [161], [163]), the group Spin($\mathcal{H}$; C_1) is closely related to the group of those canonical transformations of the CAR which are universally implementable. Sketchy indications have been given in "Sur le groupe spinoriel associé à un espace de Hilbert" (C.R. Acad. Sci. Paris, Sér. A, séance du 8 mai 1972).

Lecture Notes in Mathematics

Comprehensive leaflet on request

Vol. 111: K. H. Mayer, Relationen zwischen charakteristischen Zahlen. III, 99 Seiten. 1969. DM 16,-

Vol. 112: Colloquium on Methods of Optimization. Edited by N. N. Moiseev. IV, 293 pages. 1970. DM 18,-

Vol. 113: R. Wille, Kongruenzklassengeometrien. III, 99 Seiten. 1970. DM 16,-

Vol. 114: H. Jacquet and R. P. Langlands, Automorphic Forms on GL (2). VII, 548 pages. 1970. DM 24,-

Vol. 115: K. H. Roggenkamp and V. Huber-Dyson, Lattices over Orders I. XIX, 290 pages. 1970. DM 18,-

Vol. 116: Séminaire Pierre Lelong (Analyse) Année 1969. IV, 195 pages. 1970. DM 16,-

Vol. 117: Y. Meyer, Nombres de Pisot, Nombres de Salem et Analyse Harmonique. 63 pages. 1970. DM 16,-

Vol. 118: Proceedings of the 15th Scandinavian Congress, Oslo 1968. Edited by K. E. Aubert and W. Ljunggren. IV, 162 pages. 1970. DM 16,-

Vol. 119: M. Raynaud, Faisceaux amples sur les schémas en groupes et les espaces homogènes. III, 219 pages. 1970. DM 16,-

Vol. 120: D. Siefkes, Büchi's Monadic Second Order Successor Arithmetic. XII, 130 Seiten. 1970. DM 16,-

Vol. 121: H. S. Bear, Lectures on Gleason Parts. III, 47 pages. 1970. DM 16,-

Vol. 122: H. Zieschang, E. Vogt und H.-D. Coldewey, Flächen und ebene diskontinuierliche Gruppen. VIII, 203 Seiten. 1970. DM 16,-

Vol. 123: A. V. Jategaonkar, Left Principal Ideal Rings. VI, 145 pages. 1970. DM 16,-

Vol. 124: Séminare de Probabilités IV. Edited by P. A. Meyer. IV, 282 pages. 1970. DM 20,-

Vol. 125: Symposium on Automatic Demonstration. V, 310 pages. 1970. DM 20,-

Vol. 126: P. Schapira, Théorie des Hyperfonctions. XI, 157 pages. 1970. DM 16,-

Vol. 127: I. Stewart, Lie Algebras. IV, 97 pages. 1970. DM 16,-

Vol. 128: M. Takesaki, Tomita's Theory of Modular Hilbert Algebras and its Applications. II, 123 pages. 1970. DM 16,-

Vol. 129: K. H. Hofmann, The Duality of Compact Semigroups and C*- Bigebras. XII, 142 pages. 1970. DM 16,-

Vol. 130: F. Lorenz, Quadratische Formen über Körpern. II, 77 Seiten. 1970. DM 16,-

Vol. 131: A Borel et al., Seminar on Algebraic Groups and Related Finite Groups. VII, 321 pages. 1970. DM 22,-

Vol. 132: Symposium on Optimization. III, 348 pages. 1970. DM 22,-

Vol. 133: F. Topsøe, Topology and Measure. XIV, 79 pages. 1970. DM 16,-

Vol. 134: L. Smith, Lectures on the Eilenberg-Moore Spectral Sequence. VII, 142 pages. 1970. DM 16,-

Vol. 135: W. Stoll, Value Distribution of Holomorphic Maps into Compact Complex Manifolds. II, 267 pages. 1970. DM 18,-

Vol. 136: M. Karoubi et al., Séminaire Heidelberg-Saarbrücken-Strasbourg sur la K-Théorie. IV, 264 pages. 1970. DM 18,-

Vol. 137: Reports of the Midwest Category Seminar IV. Edited by S. MacLane. III, 139 pages. 1970. DM 16,-

Vol. 138: D. Foata et M. Schützenberger, Théorie Géométrique des Polynômes Eulériens. V, 94 pages. 1970. DM 16,-

Vol. 139: A. Badrikian, Séminaire sur les Fonctions Aléatoires Linéaires et les Mesures Cylindriques. VII, 221 pages. 1970. DM 18,-

Vol. 140: Lectures in Modern Analysis and Applications II. Edited by C. T. Taam. VI, 119 pages. 1970. DM 16,-

Vol. 141: G. Jameson, Ordered Linear Spaces. XV, 194 pages. 1970. DM 16,-

Vol. 142: K. W. Roggenkamp, Lattices over Orders II. V, 388 pages. 1970. DM 22,-

Vol. 143: K. W. Gruenberg, Cohomological Topics in Group Theory. XIV, 275 pages. 1970. DM 20,-

Vol. 144: Seminar on Differential Equations and Dynamical Systems, II. Edited by J. A. Yorke. VIII, 268 pages. 1970. DM 20,-

Vol. 145: E. J. Dubuc, Kan Extensions in Enriched Category Theory. XVI, 173 pages. 1970. DM 16,-

Vol. 146: A. B. Altman and S. Kleiman, Introduction to Grothendieck Duality Theory. II, 192 pages. 1970. DM 18,-

Vol. 147: D. E. Dobbs, Cech Cohomological Dimensions for Commutative Rings. VI, 176 pages. 1970. DM 16,-

Vol. 148: R. Azencott, Espaces de Poisson des Groupes Localement Compacts. IX, 141 pages. 1970. DM 16,-

Vol. 149: R. G. Swan and E. G. Evans, K-Theory of Finite Groups and Orders. IV, 237 pages. 1970. DM 20,-

Vol. 150: Heyer, Dualität lokalkompakter Gruppen. XIII, 372 Seiten. 1970. DM 20,-

Vol. 151: M. Demazure et A. Grothendieck, Schémas en Groupes I. (SGA 3). XV, 562 pages. 1970. DM 24,-

Vol. 152: M. Demazure et A. Grothendieck, Schémas en Groupes II. (SGA 3). IX, 654 pages. 1970. DM 24,-

Vol. 153: M. Demazure et A. Grothendieck, Schémas en Groupes III. (SGA 3). VIII, 529 pages. 1970. DM 24,-

Vol. 154: A. Lascoux et M. Berger, Variétés Kähleriennes Compactes. VII, 83 pages. 1970. DM 16,-

Vol. 155: Several Complex Variables I, Maryland 1970. Edited by J. Horváth. IV, 214 pages. 1970. DM 18,-

Vol. 156: R. Hartshorne, Ample Subvarieties of Algebraic Varieties. XIV, 256 pages. 1970. DM 20,-

Vol. 157: T. tom Dieck, K. H. Kamps und D. Puppe, Homotopietheorie. VI, 265 Seiten. 1970. DM 20,-

Vol. 158: T. G. Ostrom, Finite Translation Planes. IV. 112 pages. 1970. DM 16,-

Vol. 159: R. Ansorge und R. Hass. Konvergenz von Differenzenverfahren für lineare und nichtlineare Anfangswertaufgaben. VIII, 145 Seiten. 1970. DM 16,-

Vol. 160: L. Sucheston, Constributions to Ergodic Theory and Probability. VII, 277 pages. 1970. DM 20,-

Vol. 161: J. Stasheff, H-Spaces from a Homotopy Point of View. VI, 95 pages. 1970. DM 16,-

Vol. 162: Harish-Chandra and van Dijk, Harmonic Analysis on Reductive p-adic Groups. IV, 125 pages. 1970. DM 16,-

Vol. 163: P. Deligne, Equations Différentielles à Points Singuliers Reguliers. III, 133 pages. 1970. DM 16,-

Vol. 164: J. P. Ferrier, Seminaire sur les Algebres Complétes. II, 69 pages. 1970. DM 16,-

Vol. 165: J. M. Cohen, Stable Homotopy. V, 194 pages. 1970. DM 16,-

Vol. 166: A. J. Silberger, PGL_2 over the p-adics: its Representations, Spherical Functions, and Fourier Analysis. VII, 202 pages. 1970. DM 18,-

Vol. 167: Lavrentiev, Romanov and Vasiliev, Multidimensional Inverse Problems for Differential Equations. V, 59 pages. 1970. DM 16,-

Vol. 168: F. P. Peterson, The Steenrod Algebra and its Applications: A conference to Celebrate N. E. Steenrod's Sixtieth Birthday. VII, 317 pages. 1970. DM 22,-

Vol. 169: M. Raynaud, Anneaux Locaux Henséliens. V, 129 pages. 1970. DM 16,-

Vol. 170: Lectures in Modern Analysis and Applications III. Edited by C. T. Taam. VI, 213 pages. 1970. DM 18,-

Vol. 171: Set-Valued Mappings, Selections and Topological Properties of 2^X. Edited by W. M. Fleischman. X, 110 pages. 1970. DM 16,-

Vol. 172: Y.-T. Siu and G. Trautmann, Gap-Sheaves and Extension of Coherent Analytic Subsheaves. V, 172 pages. 1971. DM 16,-

Vol. 173: J. N. Mordeson and B. Vinograde, Structure of Arbitrary Purely Inseparable Extension Fields. IV, 138 pages. 1970. DM 16,-

Vol. 174: B. Iversen, Linear Determinants with Applications to the Picard Scheme of a Family of Algebraic Curves. VI, 69 pages. 1970. DM 16,-

Vol. 175: M. Brelot, On Topologies and Boundaries in Potential Theory. VI, 176 pages. 1971. DM 18,-

Vol. 176: H. Popp, Fundamentalgruppen algebraischer Mannigfaltigkeiten. IV, 154 Seiten. 1970. DM 16,-

Vol. 177: J. Lambek, Torsion Theories, Additive Semantics and Rings of Quotients. VI, 94 pages. 1971. DM 16,-

Please turn over

Vol. 178: Th. Bröcker und T. tom Dieck, Kobordismentheorie. XVI, 191 Seiten. 1970. DM 18,–

Vol. 179: Seminaire Bourbaki – vol. 1968/69. Exposés 347-363. IV. 295 pages. 1971. DM 22,–

Vol. 180: Séminaire Bourbaki – vol. 1969/70. Exposés 364-381. IV, 310 pages. 1971. DM 22,–

Vol. 181: F. DeMeyer and E. Ingraham, Separable Algebras over Commutative Rings. V, 157 pages. 1971. DM 16.–

Vol. 182: L. D. Baumert. Cyclic Difference Sets. VI, 166 pages. 1971. DM 16,–

Vol. 183: Analytic Theory of Differential Equations. Edited by P. F. Hsieh and A. W. J. Stoddart. VI, 225 pages. 1971. DM 20,–

Vol. 184: Symposium on Several Complex Variables, Park City, Utah, 1970. Edited by R. M. Brooks. V, 234 pages. 1971. DM 20,–

Vol. 185: Several Complex Variables II, Maryland 1970. Edited by J. Horváth. III, 287 pages. 1971. DM 24,–

Vol. 186: Recent Trends in Graph Theory. Edited by M. Capobianco/ J. B. Frechen/M. Krolik. VI, 219 pages. 1971. DM 18.–

Vol. 187: H. S. Shapiro, Topics in Approximation Theory. VIII, 275 pages. 1971. DM 22,–

Vol. 188: Symposium on Semantics of Algorithmic Languages. Edited by E. Engeler. VI, 372 pages. 1971. DM 26,–

Vol. 189: A. Weil, Dirichlet Series and Automorphic Forms. V. 164 pages. 1971. DM 16,–

Vol. 190: Martingales. A Report on a Meeting at Oberwolfach, May 17-23, 1970. Edited by H. Dinges. V, 75 pages. 1971. DM 16,–

Vol. 191: Séminaire de Probabilités V. Edited by P. A. Meyer. IV, 372 pages. 1971. DM 26,–

Vol. 192: Proceedings of Liverpool Singularities – Symposium I. Edited by C. T. C. Wall. V, 319 pages. 1971. DM 24,–

Vol. 193: Symposium on the Theory of Numerical Analysis. Edited by J. Ll. Morris. VI, 152 pages. 1971. DM 16,–

Vol. 194: M. Berger, P. Gauduchon et E. Mazet. Le Spectre d'une Variété Riemannienne. VII, 251 pages. 1971. DM 22,–

Vol. 195: Reports of the Midwest Category Seminar V. Edited by J.W. Gray and S. Mac Lane. III, 255 pages. 1971. DM 22.–

Vol. 196: H-spaces – Neuchâtel (Suisse)- Août 1970. Edited by F. Sigrist, V, 156 pages. 1971. DM 16,–

Vol. 197: Manifolds – Amsterdam 1970. Edited by N. H. Kuiper. V, 231 pages. 1971. DM 20,–

Vol. 198: M. Hervé, Analytic and Plurisubharmonic Functions in Finite and Infinite Dimensional Spaces. VI, 90 pages. 1971. DM 16.–

Vol. 199: Ch. J. Mozzochi, On the Pointwise Convergence of Fourier Series. VII, 87 pages. 1971. DM 16,–

Vol. 200: U. Neri, Singular Integrals. VII, 272 pages. 1971. DM 22,–

Vol. 201: J. H. van Lint, Coding Theory. VII, 136 pages. 1971. DM 16,–

Vol. 202: J. Benedetto, Harmonic Analysis on Totally Disconnected Sets. VIII, 261 pages. 1971. DM 22,–

Vol. 203: D. Knutson, Algebraic Spaces. VI, 261 pages. 1971. DM 22,–

Vol. 204: A. Zygmund, Intégrales Singulières. IV, 53 pages. 1971. DM 16,–

Vol. 205: Séminaire Pierre Lelong (Analyse) Année 1970. VI, 243 pages. 1971. DM 20,–

Vol. 206: Symposium on Differential Equations and Dynamical Systems. Edited by D. Chillingworth. XI, 173 pages. 1971. DM 16,–

Vol. 207: L. Bernstein, The Jacobi-Perron Algorithm – Its Theory and Application. IV, 161 pages. 1971. DM 16,–

Vol. 208: A. Grothendieck and J. P. Murre, The Tame Fundamental Group of a Formal Neighbourhood of a Divisor with Normal Crossings on a Scheme. VIII, 133 pages. 1971. DM 16,–

Vol. 209: Proceedings of Liverpool Singularities Symposium II. Edited by C. T. C. Wall. V, 280 pages. 1971. DM 22,–

Vol. 210: M. Eichler, Projective Varieties and Modular Forms. III, 118 pages. 1971. DM 16,–

Vol. 211: Théorie des Matroïdes. Édité par C. P. Bruter. III, 108 pages. 1971. DM 16,–

Vol. 212: B. Scarpellini, Proof Theory and Intuitionistic Systems. VII, 291 pages. 1971. DM 24,–

Vol. 213: H. Hogbe-Nlend, Théorie des Bornologies et Applications. V, 168 pages. 1971. DM 18,–

Vol. 214: M. Smorodinsky, Ergodic Theory, Entropy. V, 64 pages. 1971. DM 16,–

Vol. 215: P. Antonelli, D. Burghelea and P. J. Kahn, The Concordance-Homotopy Groups of Geometric Automorphism Groups. X, 140 pages. 1971. DM 16,–

Vol. 216: H. Maaß, Siegel's Modular Forms and Dirichlet Series. VII, 328 pages. 1971. DM 20,–

Vol. 217: T. J. Jech, Lectures in Set Theory with Particular Emphasis on the Method of Forcing. V, 137 pages. 1971. DM 16,–

Vol. 218: C. P. Schnorr, Zufälligkeit und Wahrscheinlichkeit. IV, 212 Seiten 1971. DM 20,–

Vol. 219: N. L. Alling and N. Greenleaf, Foundations of the Theory of Klein Surfaces. IX, 117 pages. 1971. DM 16,–

Vol. 220: W. A. Coppel, Disconjugacy. V, 148 pages. 1971. DM 16,–

Vol. 221: P. Gabriel und F. Ulmer, Lokal präsentierbare Kategorien. V, 200 Seiten. 1971. DM 18,–

Vol. 222: C. Meghea, Compactification des Espaces Harmoniques. III, 108 pages. 1971. DM 16,–

Vol. 223: U. Felgner, Models of ZF-Set Theory. VI, 173 pages. 1971. DM 16,–

Vol. 224: Revètements Etales et Groupe Fondamental. (SGA 1). Dirigé par A. Grothendieck XXII, 447 pages. 1971. DM 30,–

Vol. 225: Théorie des Intersections et Théorème de Riemann-Roch. (SGA 6). Dirigé par P. Berthelot, A. Grothendieck et L. Illusie. XII, 700 pages. 1971. DM 40,–

Vol. 226: Seminar on Potential Theory, II. Edited by H. Bauer. IV, 170 pages. 1971. DM 18,–

Vol. 227: H. L. Montgomery, Topics in Multiplicative Number Theory. IX, 178 pages. 1971. DM 18,–

Vol. 228: Conference on Applications of Numerical Analysis. Edited by J. Ll. Morris. X, 358 pages. 1971. DM 26,–

Vol. 229: J. Väisälä, Lectures on n-Dimensional Quasiconformal Mappings. XIV, 144 pages. 1971. DM 16,–

Vol. 230: L. Waelbroeck, Topological Vector Spaces and Algebras. VII, 158 pages. 1971. DM 16,–

Vol. 231: H. Reiter, L^1-Algebras and Segal Algebras. XI, 113 pages. 1971. DM 16,–

Vol. 232: T. H. Ganelius, Tauberian Remainder Theorems. VI, 75 pages. 1971. DM 16,–

Vol. 233: C. P. Tsokos and W. J. Padgett. Random Integral Equations with Applications to Stochastic Systems. VII, 174 pages. 1971. DM 18,–

Vol. 234: A. Andreotti and W. Stoll. Analytic and Algebraic Dependence of Meromorphic Functions. III, 390 pages. 1971. DM 26,–

Vol. 235: Global Differentiable Dynamics. Edited by O. Hájek, A. J. Lohwater, and R. McCann. X, 140 pages. 1971. DM 16,–

Vol. 236: M. Barr, P. A. Grillet, and D. H. van Osdol. Exact Categories and Categories of Sheaves. VII, 239 pages. 1971, DM 20,–

Vol. 237: B. Stenström. Rings and Modules of Quotients. VII, 136 pages. 1971. DM 16,–

Vol. 238: Der kanonische Modul eines Cohen-Macaulay-Rings. Herausgegeben von Jürgen Herzog und Ernst Kunz. VI, 103 Seiten. 1971. DM 16,–

Vol. 239: L. Illusie, Complexe Cotangent et Déformations I. XV, 355 pages. 1971. DM 26,–

Vol. 240: A. Kerber, Representations of Permutation Groups I. VII, 192 pages. 1971. DM 18,–

Vol. 241: S. Kaneyuki, Homogeneous Bounded Domains and Siegel Domains. V, 89 pages. 1971. DM 16,–

Vol. 242: R. R. Coifman et G. Weiss, Analyse Harmonique Non-Commutative sur Certains Espaces. V, 160 pages. 1971. DM 16,–

Vol. 243: Japan-United States Seminar on Ordinary Differential and Functional Equations. Edited by M. Urabe. VIII, 332 pages. 1971. DM 26,–

Vol. 244: Séminaire Bourbaki – vol. 1970/71. Exposés 382–399. IV, 356 pages. 1971. DM 26,–

Vol. 245: D. E. Cohen, Groups of Cohomological Dimension One. V, 99 pages. 1972. DM 16,–